History of
Japan's Intrusions
into Korea

Seeing Dokdo Through 30 Images and Historical Documents

● Edited by Dokdo Research Institute
Northeast Asian History Foundation

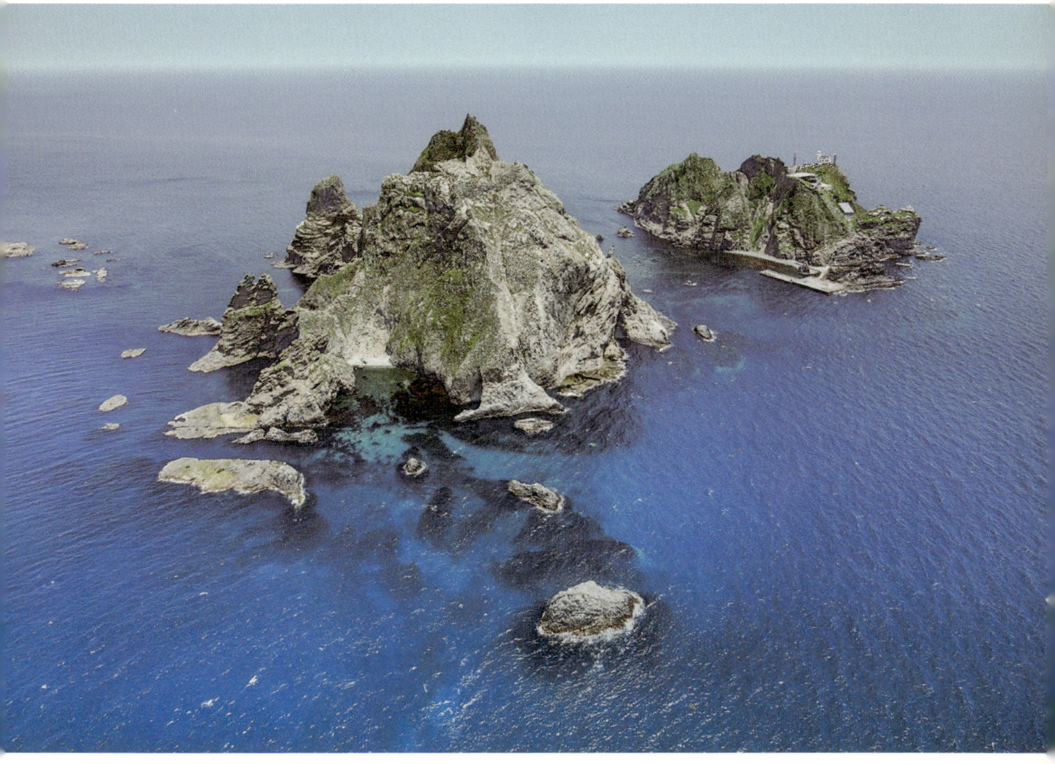

Image of Dokdo (ⓒ KimJungman)

Preface

Koreans commonly refer to Japan as the "near yet distant country." Geographical proximity between Korea and Japan necessitates that Korea and Japan be close neighbors, yet throughout history, relations between the two countries have always been "distant." Conflicts have regularly arisen due to "historical problems." Unfortunately, ever since Korea and Japan normalized diplomatic relations in 1965, current relations between the two countries are at a nadir.

The Korean-Japanese conflict is rooted in 'history problems.' This is due to a neglect of these problems during the normalization of diplomatic relations in 1965. There have been modest efforts in subsequent years to assuage that "wound" and become "close countries." However, problems recently became worse as Japan tries to deepen the wound. Japan's claim of territorial sovereignty over Dokdo can also be understood within this very context.

To foster "neighborly" relations between the two countries and build a peaceful system of mutual coexistence, it is crucial for both Korea and Japan to correctly view history instead of neglecting a "history of wounds and resentment." Therefore, our foundation intends to publish "Understanding Japan's History of Aggression," a series of books which citizens and students can easily read and comprehend. Many facts in these books will be very familiar thanks to Korea's excellent history curriculum, but we intend to analyze and interpret them in a new light. We included many historical documents and images to help readers easily understand issues with greater lucidity than most general history books.

This book is about Dokdo as Korea's territory. Dokdo was the first Korean territory that the Japanese robbed during her invasion of the Korean Empire.

In 1905, as the Russo-Japanese War was still ongoing, Japan unilaterally declared that Dokdo was "her territory." *Seeing Dokdo Through 30 Images and Historical Documents* compiled essential images and documents which historically, geographically, and legally belong to Korea under international law.

From Ulleungdo, anyone can observe Dokdo for 40 days every year. There are also records and photographs of Dokdo. That one can see with their naked eyes also implies Dokdo was always a part of its residents' daily lives. Dokdo is literally a space in which people can lead their lives. In addition, there are numerous ancient records confirming that Dokdo(Usando) is Korean territory. *A History of the Three Kingdoms*, *A History of Goryeo*, *The Royal Annals of King Sejong*, *A Geographical Survey of Korea*, *An Encyclopedia of Korean Customs and Culture*, and various maps are but a few examples. Moreover, there are many Japanese documents and maps recording that Dokdo is not a part of Japanese territory. This book contains 30 images of historical documents and maps related to such records. It also contains current images of Dokdo under Korean control.

Our foundation intends to continue discovering historical data confirming that Dokdo is Korean territory and publicizing this fact in Korea and abroad. We hope that *Seeing Dokdo Through 30 Images and Historical Documents* offers an opportunity to reflect on the importance of Dokdo as Korea's territory.

Kim Dohyung,
President, Northeast Asian History Foundation,
September 2020

Contents

Preface • 2

1. Dokdo as Seen From Ulleungdo • 9

Ancient Records
2. Kim Bu-sik, *A History of the Three Kingdoms* • 15
3. *A Geographical Survey of Korea* in *The Royal Annals of King Sejong* • 17
4. *A Comprehensive Map of the Eight Provinces* from *A New and an Expanded Encyclopedia of Korean Geography* • 21
5. Saito Toyonobu, *A Record of General Observations on Oki Island* • 23

The An Yong-bok Incident and the Advancement of Korean and Japanese Perceptions About Dokdo
6. *A Book About a Korean Boat Which Arrived on the Coast in 1696* • 29
7. *The Tottori Han's Reply to the Bakufu's Question* • 33
8. Jang Han-sang and *A Record of Incidents on Ulleungdo* • 37

The Spread of Geographical Awareness About Ulleungdo and Dokdo
9. Frenchman D' Anville's *A Complete Map of the Korean Kingdom* • 45
10. *The Grand Map of Korea* • 49
11. *An Encyclopedia of Korean Customs and Culture* • 53
12. Hayashi Shihei, *A General Cartographic Record of Japan* in *An Illustrated Description of Three Countries* • 55
13. *A Comprehensive Map of Our Country* • 59

Information on Dokdo from 19th Century Maps and Documents
14. *Kangwon Province* in *Haedongjeondo (A Complete Map of Korea)* • 65
15. Kim Dae-geon, *A Complete Map of Korea* (1845) • 69
16. Russia, *A Map of Korea's East Sea Coastlines* • 71
17. *Haejwajeondo (A Complete Map of Korea)* • 77
18. *An Intelligence Report on the Internal Affairs of Korea* • 81
19. Japanese Ministry of the Navy, *A Map of Korea's East Coast* • 87
20. *Dajokan Directive* and *A Map of Isotakeshima* • 91
21. The Japanese Army, *A Complete Map of Great Japan* • 97

Japan's Intrusion and the Korean Empire's Resolute Management of Territories
22. Lee Kyu-won, *An Investigative Diary on Ulleungdo* • 103
23. Imperial Decree No. 41 of the Korean Empire • 107
24. The Russo-Japanese War and the Dokdo Watchtower • 113
25. Acting Provincial Governor, Kangwon Province, *The Lee Myeong-rae Report* • 119

The Defeat of Japan in World War II and the Return of Dokdo to Korea
26. SCAPIN No. 677 and Related Maps • 127
27. Investigation Commission for Scholarly Research on Ulleungdo and Dokdo • 133
28. The Dokdo Bombing Incident (1948, 1952) • 141
29. The Declaration of a Line of Peace, Gazette, and Map(1952) • 145
30. The Dokdo Volunteer Garrison and the Dokdo Security Police • 149

Index • 152

List of Primary Sources

Kkak-kae Mound, Dodong, Ulleungdo(2008. 11. 22) • 9
Saekak Mound, Sadong, Ulleungdo(2008. 11. 22) • 9
Distance from Ulleungdo and Oki Island • 10
A History of the Three Kingdoms • 15
A Geographical Survey of Korea in The Royal Annals of King Sejong • 17
A Comprehensive Map of the Eight Provinces from A New and an Expanded Encyclopedia of Korean Geography • 21
A Record of General Observations on Oki Island • 23
A Book About a Korean boat Which Arrived on the Coast in 1696 • 29
The Tottori Han's Reply to the Bakufu's Question • 33
A Record of Incidents on Ulleungdo • 37
A Complete Map of the Korean Kingdom • 45
Enlarged Portion of A Complete Map of the Korean Kungdom Showing Ulleungdo and Dokdo • 46
The Grand Map of Korea • 49
Ulleungdo and Usando in The Grand Map of Korea • 50
An Encyclopedia of Korean Customs and Culture • 53
A General Cartographic Record of Japan in An Illustrated Description of Three Countries • 55
A Comprehensive Map of Our Country • 59
Haedongjeondo (A Complete Map of Korea) • 65
A Complete Map of Korea • 69
A Map of Korea's East Sea Coastlines • 71
Haejwajeondo (A Complete Map of Korea) • 77
Magnified Portion of Haejwajeondo (A Complete Map of Korea) • 78
An Intelligence Report on the Internal Affairs of Korea • 81
A Map of Korea's East Coast • 87
Dajokan Directive • 91
Map of Isotakeshima • 93
A Complete Map of Great Japan • 97
An Investigative Diary on Ulleungdo • 103

Imperial Decree No. 41 of the Korean Empire • 107
Dokdo Watchtower • 113
The Lee Myeong-rae Report • 119
SCAPIN No. 677 and Related Maps • 127
Investigation Commission for Scholarly Research on Ulleungdo and Dokdo(1947.8) • 133
Investigation Commission for Scholarly Research on Ulleungdo and Dokdo(1953) • 133
A Panoramic view of West Island as seen from East Island's Mongdol Coast,
    site of the Dokdo granite plaque • 139
Memorial Monument for Victimized Fishermen • 141
Dong-a Il-bo(East Asia Daily) article of September 21, 1952 • 143
A Public Notice Gazette Containing the Declaration of a Line of Peace • 145
Map Accompanying the Gazette • 145
The Dokdo Volunteer Garrison inspecting with a large telescope • 149
The Dokdo Security Police Raising the Republic of Korea's National Flag • 150
Notation of Dokdo as a Territory of the Republic of Korea • 158
The Dokdo Security Police • 158

〈Supplemental Source〉

〈Supplemental Source 1〉 Excerpt from the *Royal Annals of King Taejong* • 20
〈Supplemental Source 2〉 *An Order Prohibiting the Crossing of the Seas* • 36
〈Supplemental Source 3〉 *A Complete Map of the Korean Kingdom* :
    The Extension of the Régis Line • 48
〈Supplemental Source 4〉 Wooden Fences Marking the Border of the Restricted Area
    ("Willow-Fence Border") in *A Complete Map of the Korean Kingdom* • 48
〈Supplemental Source 5〉 *A Complete Map of the Korean Nation* from *An Illustrated
    Description of Three Countries* • 57
〈Supplemental Source 6〉 *A Grand Map of Three Countries—Hokkaido, Korea, and Japan* • 58
〈Supplemental Source 7〉 *A Gazetteer on Korea's Waterways*(1899) • 89
〈Supplemental Source 8〉 *A Gazetteer on Korea's Waterways*(1907) • 90
〈Supplemental Source 9〉 Inscription Left on Rocks at Hakpo, Tae-ha Village, Ulleungdo • 106
〈Supplemental Source 10〉 A Request from the Minister of Internal Affairs (1900) • 111
〈Supplemental Source 11〉 *Dae-han Maeil Shinbo (The Dae-han Daily News)* (May 1, 1906) • 122
〈Supplemental Source 12〉 The Front and Back of the Plaque Erected in 1953 • 138
〈Supplemental Source 13〉 The plaque, reinstalled on July 6, 2015 • 139

1

Dokdo as Seen From Ulleungdo

Dokdo was Linked with Ulleungdo from Ancient Times

Kkak-kae Mound, Dodong, Ulleungdo(2008. 11. 22) (Source: The Northeast Asian History Foundation)

Saekak Mound, Sadong, Ulleungdo(2008. 11. 22) (Source: The Northeast Asian History Foundation)

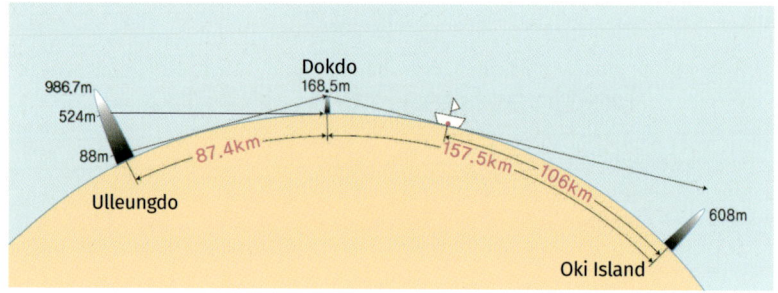

Distance from Ulleungdo and Oki Island(Source: The Northeast Asian History Foundation)

On a clear day, Dokdo is visible from Ulleungdo with the naked eye. Thus, Ulleungdo's residents naturally regarded Dokdo as Korean territory. The fact that one can see Dokdo from Ulleungdo is proven in *A Geographical Survey of Korea* in *The Royal Annals of King Sejong* (1454) or *A Record of Incidents in Ulleungdo* (to be explained in greater detail in subsequent sections). In particular, *A Record of Incidents in Ulleungdo*, which Jang Han-sang, Provincial Magistrate of Samcheok, wrote when he was commissioned as an Inspector to Ulleungdo, records that he was able to view Dokdo from Seongin Peak. That one can view Dokdo from Ulleungdo is an immutable fact.

The Northeast Asian History Foundation conducted an investigation on observing Dokdo with the naked eye (recording the number of days one can physically see Dokdo) and organized the results using photographs and records. The results were compiled over the course of a year and six months, and during that period, Dokdo was visible with the naked eye for 56 days. Upon examining the photographs, one immediately notices that there is a photograph of Dokdo taken with a power pole standing in the front yard of a house in Ulleungdo as its background. Another photograph of Dokdo is taken against the background of a persimmon tree blossoming in the front yard of a farmhouse. In short, Ulleungdo's residents naturally consider Dokdo to be a part of daily life.

There was a moment when Japan wished to deny that Dokdo could be observed from Ulleungdo. Japan believed that if she could confirm that Dokdo was unobservable from Ulleungdo, it could be used to prove that Koreans were not aware of Dokdo's existence. However, that Dokdo can be observed from Ulleungdo remains an indisputable fact.

By contrast, it is essentially impossible to see Dokdo from Oki Island, which is near Japan. The distance from Dokdo to Oki Island is 1.8 times the distance between Dokdo and Ulleungdo. Among islands near Dokdo, Ulleungdo is practically the only island from which Dokdo is observable. Consequently, even Japanese documents use expressions such as "Dokdo as part of Ulleungdo," "Dokdo near Ulleungdo," and "Ulleungdo and the Other Island (Dokdo)."* The fact that Dokdo can be seen from Ulleungdo not only suggests that Koreans were long cognizant of Dokdo's existence but also that Ulleungdo and Dokdo were, in terms of historical geography, considered to be closely related. Dokdo and Ulleungdo's proximity was such that the former was regarded as a constituent of the latter, respectively as a small and a large island.

---

* Japan currently calls Dokdo, "Takeshima," (竹島) but Japanese records before the 20th century refer to Dokdo as Matsushima (松島) and Ulleungdo as Takeshima (竹島).

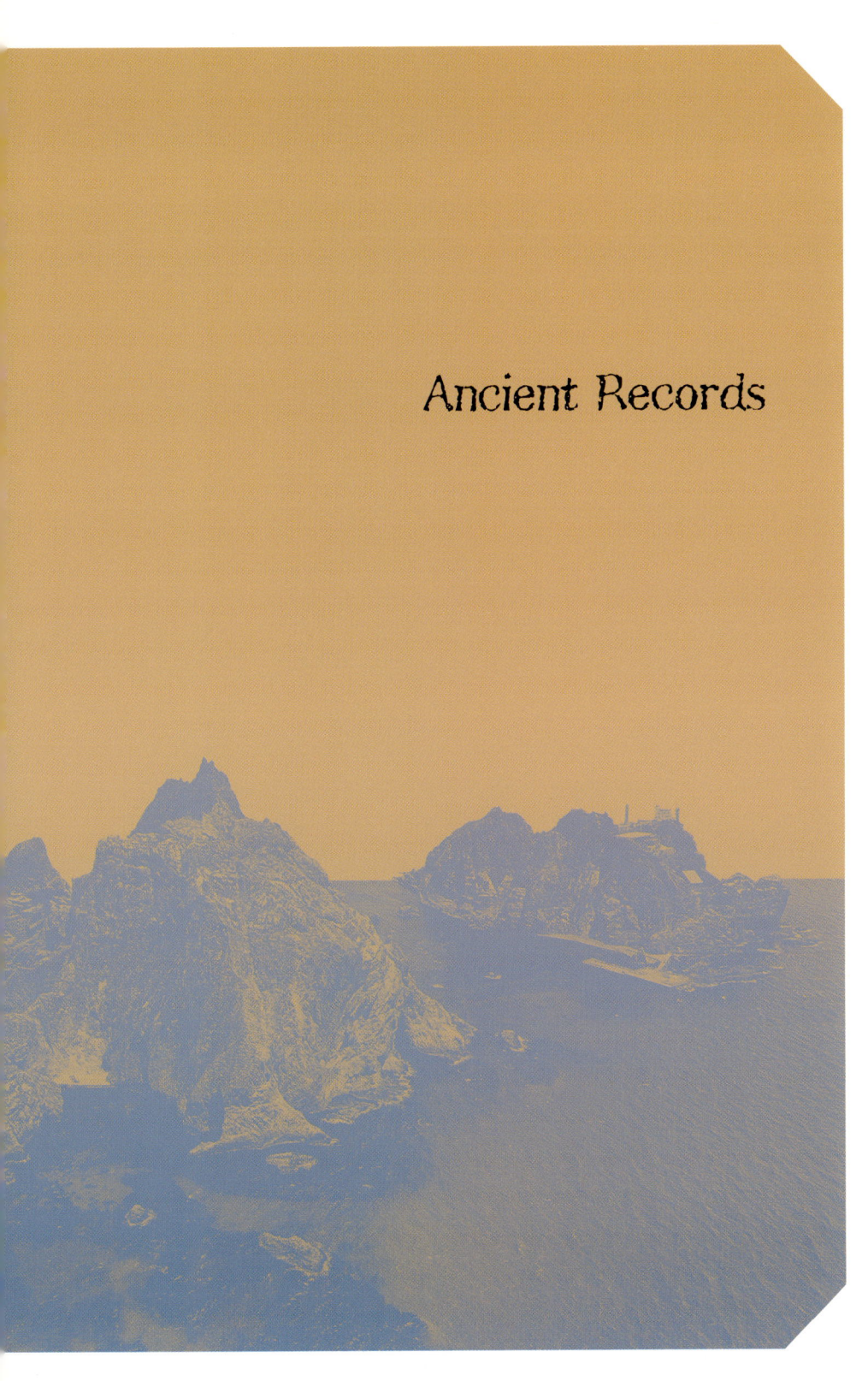
Ancient Records

2

Kim Bu-sik, *A History of the Three Kingdoms*

Silla Incorporates Usankuk (Ulleungdo and Dokdo)

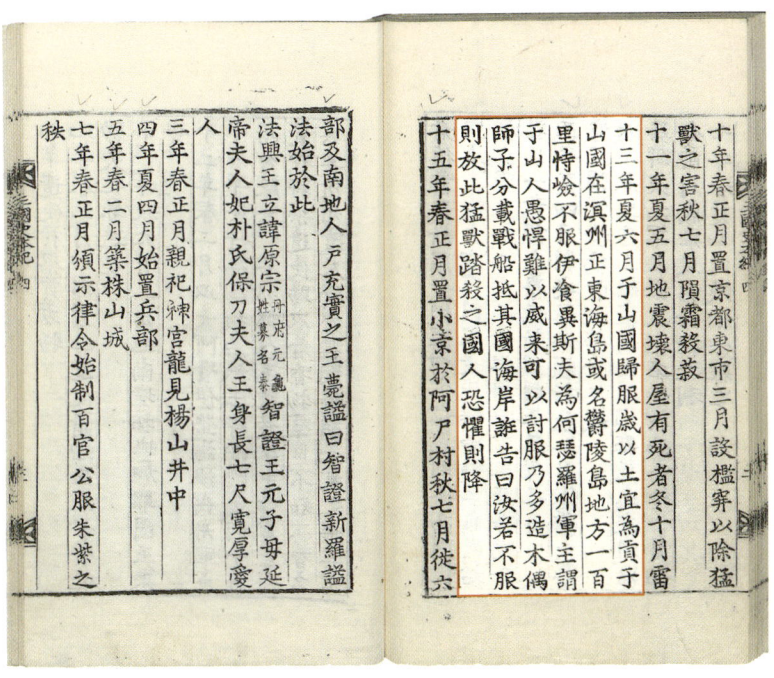

A History of the Three Kingdoms (Source: Kyujanggak Institute for Korean Studies)

In the 13th Year of the reign of King Jijeung, Usankuk surrendered and submitted, promising to offer local products annually as tribute. Usankuk is an island located directly east of Myeongju, or what is now known as Gangneung, and is also known as Ulleungdo. The island's parameters are about 100 Li. The island's inhabitants initially relied on the rough terrain and refused to surrender. When Lee Sa-bu became Military Commander of Hasla Province (as Myeongju was called during the Silla Period), he said that since Usankuk's people were "blockish and wild," they could not be subdued by brute force, and had to be forced to capitulate by the use of strategy. He promptly ordered his men to carve lions out of wood, place them aboard individual naval ships, and arrived at the coast of Usankuk. He lied that he would "unleash these beasts and let them trample you to death" to Usankuk's residents, who immediately surrendered in fear.

A History of the Three Kingdoms (1145) is a book published during the Goryeo period by Kim Bu-sik. It is the oldest history book among those still extant in Korea. Among the books still extant, *The Main Historical Record of Silla* notes that Usankuk(于山國) became a part of Silla after it surrendered during the 13th Year of King Jijeung's reign (512). Ancient Usankuk was a tribal, semi-agrarian and semi-fishing society. Its territory was comprised of Dokdo within visually verifiable distance, as well as Ulleungdo and its constituent islands. Thus, Dokdo became a constituent island of Ulleungdo after surrendering to Silla in 512 B.C.E., whereupon it became part of the Korean Peninsula's historical and cultural domain. *A History of the Three Kingdoms* separately records Usankuk as the name of a nation and Ulleungdo as the name of an island. The records of *A History of the Three Kingdoms* were passed down and Ulleungdo and Dokdo were also recorded in *A History of Goryeo* (1451), *The Annals of King Sejong* (1454), and *A New and an Expanded Encyclopedia of Korean Geography* (1481).

3

A Geographical Survey of Korea in *The Royal Annals of King Sejong*

Ulleungdo and Dokdo Were Part of Uljin County During the Early Joseon Period

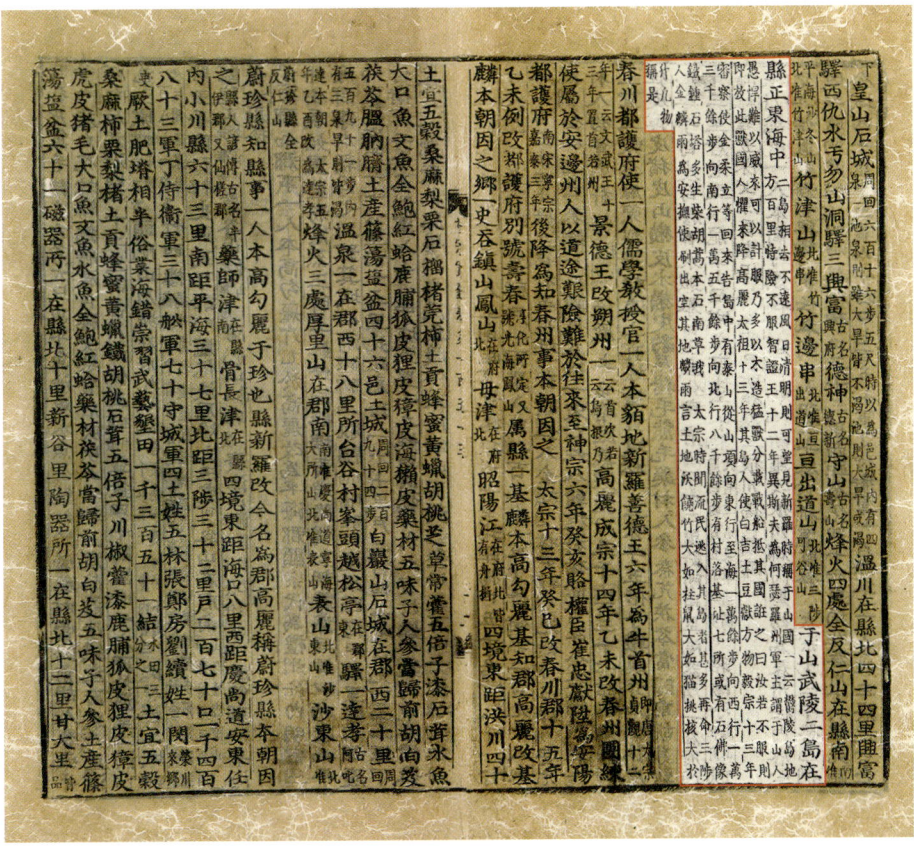

A Geographical Survey of Korea in *The Royal Annals of King Sejong*
(Source: Kyujanggak Institute for Korean Studies)

Usan and Mureung; the two islands are located in the middle of the ocean, directly east of Uljin County.

Since the distance between the two islands is not far, it is possible to see them from afar on a clear day. The two islands were called Usankuk during the Silla Period and were also collectively referred to as "Ulleungdo." They have a parameter of about a 100 Li, and the Usankuk residents would not capitulate to invasion because they trusted the rough terrain. Lee Sa-bu, who became Military Commander of Hasla Province in the 12th Year of King Jijeung's reign (511) said that because the residents of Usankuk were "foolish and wild," a display of majesty alone would not guarantee their surrender, and therefore, the employment of a strategy would subjugate them. Therefore, he ordered that many wild beasts be carved out of wood, and loaded them on multiple naval ships and reached Usankuk, whereupon he deceived its residents with these words. "Should you not surrender, I shall unleash these beasts upon you." The apprehensive residents surrendered to Lee Sa-bu.

During the 13th Year in the reign of the Goryeo Dynasty's Taejong, the island's residents sent Baekgil and Todu to pay tribute. During the 13th Year in the reign of Eui-jong, Royal Examiner Kim Yoo-rip and others returned from Ulleungdo and reported the following:

"There is a large mountain in the middle of the island, and it takes about 10,000 steps eastward from the mountain's peak to the sea; about 13,000 steps westward to the sea; about 15,000 steps southward to the sea and 8,000 steps northward towards the sea. There are about 7 village settlements, several stone statues of Buddha, metal bells, and stone pagodas, as well as numerous Chinese thoroughwax, slender angelica and Seok-nam-cho (literally, "southern rock plant," a native plant of Ulleungdo)."

During the reign of the Joseon Dynasty's Taejong (in the actual text, the name incorrectly appears as "Taejo." Refer to Supplemental Source 1.), the court received reports informing that many "migrants" fled towards or entered the island in large numbers, and ordered Public Security Manager Kim In-woo from Samcheok to thoroughly search for the "migrants" and immediately vacate the island. Kim reported that "bamboo trees are thick as pillars, mice are almost as large as cats, and peach seeds are the size of gourds, and many other objects are of such scale."

A Geographical Survey of Korea(1454), which is incorporated in the *Annals of King Sejong*, categorizes Usan and Mureung as part of Uljin County. Usan and Mureung appear as entries, and the main text, which provides general information about Usan and Mureung, is written in a large font. explanatory notes explaining the specific histories of these islands appear in a small font underneath the main text. This record is an important piece of historical evidence which, firstly, shows Usan and Mureung as two islands in the middle of the East Sea. This confirms that Ulleungdo is one island by classifying it as a single entry, which is an extension from the explanation given in *A History of Goryeo*'s version of *A Geographical Survey of Korea*, which had introduced the myth that Ulleungdo was comprised of two islands. Secondly, the record notes that the two islands are not very distant from one another and therefore, both could be viewed from afar. Based on these two pieces of evidence, we can know that Usan and Mureung respectively refer to Dokdo and Ulleungdo.

The geographical knowledge from *The Annals of King Sejong*'s version of *A Geographical Survey of Korea* gets passed on to *A New and an Extended Encyclopedia of Korean Geography*, published during the reign of King Seongjong, which categorizes Usando and Ulleungdo as part of Uljin County.

⟨**Supplemental Source 1**⟩ Excerpt from the *Royal Annals of King Taejong*

Kim In-woo became Public Security Manager of Mureung and Adjacent Regions. Minister of Finance Park Seup reported, "Your Highness, when your humble servant once served as Public Observer of Kangwon Province, I heard that Mureung Island has a parameter of about 210 Li, and there is a small island adjacent to it, which is only about 500,000 square meters wide. The main road is only wide enough for one person to pass through at once, and two people may not walk side by side. There was once a man by the name of Bang Ji-yong, who was the head of 15 families and occasionally committed larceny by posing as a Wakou pirate. Since there is a man residing in Samcheok who is very familiar with the island, why not send him to observe what is happening on the island?" The king approved of the advice and summoned Kim In-woo a former Head Officer of Provincial Security and inquired about Mureung Island. Kim In-woo reported, 'since Mureung Island is far out in the middle of the sea, very few people reside or visit, and therefore, there are those who escape to the island to evade military service. Should too many people live on the island, the Japanese will most certainly and eventually return to the island, commit larceny and ultimately, invade Kangwon Province.' The king considered this advice to be right and proceeded to name Kim In-woo, Public Security Manager of Mureung and Adjacent Regions, and Lee Man as Kim's attendant. The king provided two naval ships, two oarsmen, two shipmates, cannons, gunpowder, and provisions, instructing Kim to go to the island and inform the head officer of the island to come to court, and gave Kim and Lee apparel, hats and shoes.

September 12, Sixteenth Year of King Taejong's Reign

4

A Comprehensive Map of the Eight Provinces from *A New and an Expanded Encyclopedia of Korean Geography*

The Two Islands Begin to Appear on Korean Maps

*A Comprehensive Map of the Eight Provinces fro*m *A New and an Expanded Encyclopedia of Korean Geography*(27×34.2cm) (Source: Seoul Museum of History)

A Comprehensive Map of the Eight Provinces is Korea's first map which records Dokdo (Usando). *A Comprehensive Map of Korea's Eight Provinces* is one of the nine maps collected in *A New and an Expanded Encyclopedia of Korean Geography*. The nine maps are of Kyeonggi, Chungcheong, Kyongsang, Jeolla, Hwanghae, Kangwon, Hamkyeong, Pyeong-an Provinces, and *A Comprehensive Map of Korea's Eight Provinces*. *A Comprehensive Map of Korea's Eight Provinces* is a map which depicts all of Korea's eight provinces.

A New and an Expanded Encyclopedia of Korean Geography is a new and updated edition of *A Geographical Encyclopedia of Korea*. Both books served as main Geography textbooks during the Joseon Period, and Japan also adopted *A New and Extended Geographical Encyclopedia* as the standard guide to Korea's geography. *A Comprehensive Map of the Eight Provinces* records Usando and Ulleungdo in the middle of the East Sea. The map's depiction of Usando and Ulleungdo in the East Sea clearly represents the geographical perception of the two islands during the Joseon Period. Nevertheless, that Usando is drawn to the West rather than East of Ulleungdo suggests the lack of precision in surveying techniques, which caused a considerable reliance on pre-existing records to draw the two islands.

5

Saito Toyonobu, *A Record of General Observations on Oki Island*

Even During This Period,
Ulleungdo and Dokdo Were Outside Japan's Border

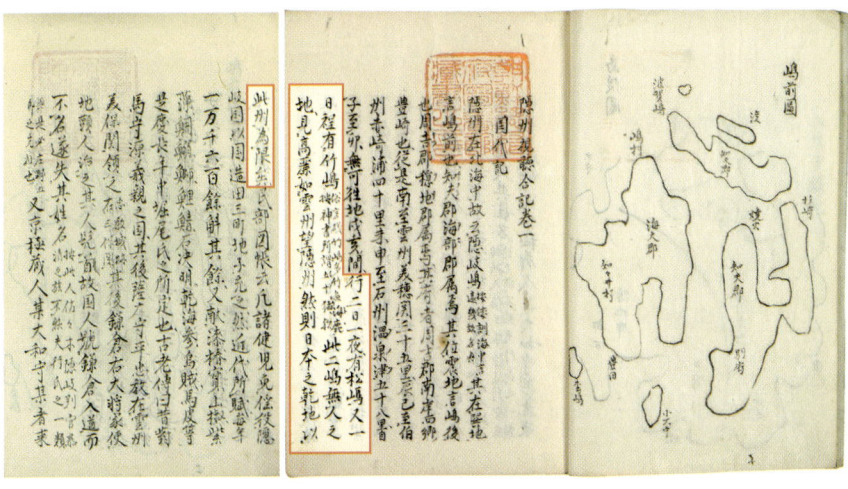

A Record of General Observations on Oki Island (Source: National Library of Korea)

If one travels northwest for two days, there is Matsushima (Dokdo), and if one travels further for another day, there is Takeshima (Ulleungdo) (Commonly referred to as "Isotakeshima (Ulleungdo)," the island harbors diverse species of bamboo, fish, and sea lion). There are no residents on these islands, and observing Korea from them is akin to viewing Oki Island from Shimane Prefecture. Therefore, this island serves as the northwestern limit of Japan's territories.

A Record of General Observations on Oki Island is a Japanese geography text which records the history and geography of Oki Island. Saito Toyonobu, the *Hanshi* of Matsue, recorded incidents he had either directly observed or heard during his patrols of Oki Island. Book 1, titled "Records in the Age of the Nation," which comprises the book's general thesis, records his observations about Dokdo and Ulleungdo. He recorded that if one travels northwest for two days from Oki Island, there is Matsushima (Dokdo); if one travels further for another day, there is Takeshima (Ulleungdo), and that these were two uninhabited islands adjacent to Korea. Saito then records that Oki Island marks the northwestern limits of Japan's territory. In short, this document proves that the Japanese, by their own volition, excluded Dokdo and Ulleungdo from their territories.

Japan had once presented *A Record of General Observations on Oki Island* as evidence to claim ownership over Dokdo (Opinion issued from the Japanese government on September 25, 1954). According to the Japanese, the text mentions Ulleungdo and Dokdo and declares them to comprise the northwestern border of Japan. Subsequently, there have been debates on Japan's claim regarding whether "this island" from the sentence, "Therefore, this island serves as the northwestern limit of Japan's territories," refers to Takeshima (Ulleungdo) or Oki Island.

However, the Japanese government no longer presents *A Record of General Observations on Oki Island* as evidence to claim sovereignty over Dokdo, for many Korean and Japanese scholars have clearly determined that "this island" refers to Oki Island. *A Record of General Observations on Oki Island*'s notation of Oki Island as the northwestern limit of Japan's territory proves that Japan also recognized Ulleungdo and Dokdo as Korean territory during the 17th century.

The An Yong-bok Incident and the Advancement of Korean and Japanese Perceptions About Dokdo

6

A Book About a Korean Boat Which Arrived on the Coast in 1696

An Investigative Report About a Korean Boat Which Had Arrived on Japan's Coast

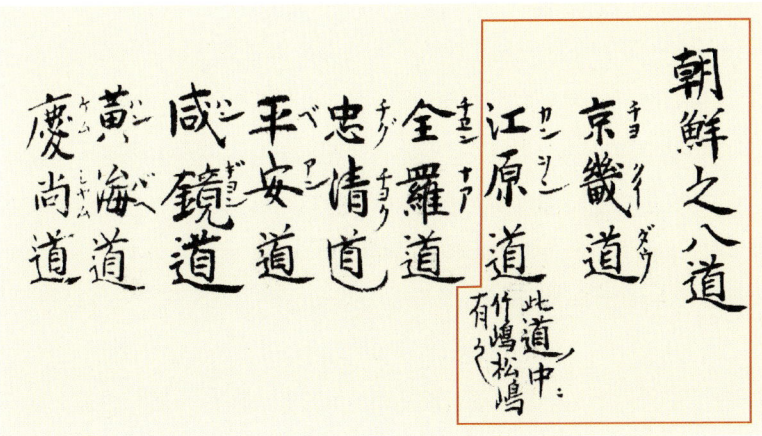

A Book About a Korean boat Which Arrived on the Coast in 1696 (Source: Murakami Sukekuro)

8 Provinces of Korea

(...)

Kangwon-do

Takeshima and Matsushima are part of this province.

A Book About a Korean boat Which Arrived on the Coast in 1696 (hereafter shortened to *A Book of 1696*) is a document which records that the map in An Yong-bok's possession when he crossed the sea from Korea to Japan for the second time noted Dokdo and Ulleungdo as Korean territories.

During the 19th Year in the Reign of King Sukjong (1693), Dongrae's An Yong-bok, who fished off the waters of Ulleungdo for a living, clashed with Japanese fishermen over fishing rights in the area, and, along with Park Eo-dun, got kidnapped to Japan. This incident induced what later became known as the "Ulleungdo Controversy." (Confer Source 7) An Yong-bok declared to the Viceroy of Hoki that Ulleungdo is Korean territory. The Viceroy, in compliance with an order from the Bakufu, transported An to Nagasaki and sent him back to Busan via Tsushima Island. An returned to Korea in nine months.

An Yong-bok and his cohort went to Japan for the second time in May 1696. Despite the Bakufu's announcement forbidding Japanese fishermen from crossing the waters between Korea and Japan, the lord of Tsushima's postponement of the decree's registration delayed the implementation of the Bakufu's order. Many Japanese were still fishing near the waters of Ulleungdo, and An, deciding to take matters into his own hands, carried out his plan to cross over to Japan.

A Book of 1696 is a document that an official of Oki Island wrote while he was investigating An and his cohort as they were visiting Oki Island as part of their trip to Tottori Prefecture. The full title *of A Book of 1696* means a book about a ship which arrived on Oki Island's coast in 1696. The book records An and his cohort's manner of dress, objective behind crossing the sea, and incidents on Oki Island, along with a map of Korea attached at the end of the book, which the photograph at the beginning of this chapter shows.

An possessed a map of Korea organized into eight separate sheets, which he submitted when he was being investigated on Oki Island. The official of Oki Island transcribed "Kyeonggido" and the Korean names of the other provinces as a list and transliterated the Korean pronunciations of the provinces in Japanese. This became known as *A List of Korea's Eight Provinces*. It is remarkable that the official wrote "Takeshima and Matsushima are part of this province" underneath "Kangwon Province." During the investigation, An Yong-bok argued that what the Japanese call Takeshima and Matsushima are Ulleungdo and Dokdo, and that the map of Korea in his possession also records the two islands as part of Kangwon Province. *A Book of 1696* records An as testifying that Matsushima is noted as "Jasando(子山島)," but this is an incorrect rendering of "Usando(于山島)" in Chinese characters, and actually refers to Dokdo.

Upon examining ancient records such as *An Encyclopedia of Korean Customs and Culture*, one finds statements such as "Usando is what the Japanese refer to as Matsushima," which suggest that what An Yong-bok referred to as "Jasando" is actually Usando and Matsushima is what the

Japanese called Dokdo in the 17th century and refers to Usando.

A List of Korea's Eight Provinces shows that the purpose behind An Yong-bok's visit to Tottori Prefecture during his second crossing to Japan was to affirm that Ulleungdo and Dokdo were part of Kangwon Province and Korean territory and that Ulleungdo and Dokdo were noted as Korean territory on Korean maps.

7
The Tottori Han's Reply to the Bakufu's Question
"Ulleungdo and Dokdo are not part of Tottori Han."

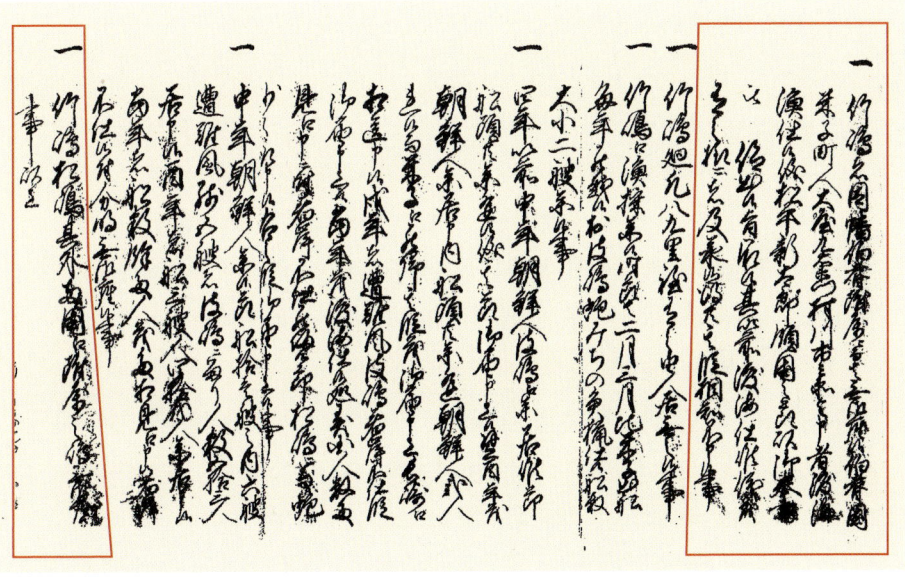

The Tottori Han's Reply to the Bakufu's Question (Source: Tottori Prefectural Museum)

Article 1: Takeshima is not part of Inaba and Hoki (Tottori Prefecture). Oya Kyuemon and Murakawa Ichibe of Yonago, Hoki crossed the sea and engaged in fishing. When Matsudaira Shintaro was on patrolling the area, he gave instructions via a sealed letter to cross the sea. We are aware of past crossings, but we are not entirely familiar with specific details.

Article 6: In addition to Takeshima (Ulleungdo) and Matsushima (Dokdo), there are no other islands which are part of Inaba and Hoki.

This report of statements not only reflects Tottori-han's opinion that Dokdo is not Japanese territory, but also that Dokdo is included in the Bakufu's ban on crossing the sea between Korea and Japan. Disputes between the Edo Bakufu and the Korean government on whether Ulleungdo is Korean or Japanese territory began in December 1693. These disputes are known as the "Ulleungdo Controversy" in Korea and the "Takeshima Incident" in Japan. The disputes began when An Yong-bok and Park Eo-dun, who were among Korean fishermen who went to Japan, got kidnapped by Japanese fishermen, and were ultimately concluded when the Bakufu imposed a ban on crossing the sea between Korea and Japan in 1699. "The Tottori Han's Reply to the Bakufu's Question" served as a basis for the Bakufu's decision. On December 24, 1695, the Bakufu sent a questionnaire to the Tottori Han, and the photograph shows the document which contains the contents of the Tottori Han's reply while the "Ulleungdo Controversy" was still ongoing. The first Article of the Bakufu's questionnaire was "Since when was Ulleungdo part of Hoki and Inaba (two regions located in Tottori Prefecture)?" to which the Tottori Han gave the reply as shown. The Bakufu proceeded to ask in Article 6, "Are there other islands besides Ulleungdo which are part of Inaba and Hoki, and have there been individuals from these areas who engaged in fishing?" to which Tottori Han gave the reply as shown.*

Of the six replies, replies to Questions 1-5 were about Takeshima (Ulleungdo), which the Bakufu had inquired about, and in its reply to Question 6, the Tottori Han included Dokdo by responding, "There are no additional islands, including Takeshima(Ulleungdo) and Matsushima(Dokdo), which are part of the two regions." The Edo Bakufu imposed a ban on fishermen, which prohibited them from crossing the sea between Korea and Japan(See Supplemental Source 2). It is important to note that, in spite of the Bakufu not inquiring about Matsushima(Dokdo), the Tottori Han included it along with Takeshima(Ulleungdo) and declared them as not being part of Tottori Prefecture. This suggests that the ban on crossing the sea not only applied to Takeshima(Ulleungdo) but Matsushima(Dokdo) as well.

---

* The reply consisted of six Articles and the other four Articles were the following: Article 2 responded that Ulleungdo's parameter is about 80-90 Li and there were no inhabitants. Article 3 responded that 2~3 ships head to Ulleungdo every February and March, and that abalones and sea lions were frequently caught. Article 4 briefly describes Ulleungdo and notes that although there have been visits to Ulleungdo, Japanese ships could not anchor there because of Korean residents and hence, had to stop by Dokdo to catch abalones. Article 5 notes that there were at least 40-50 people who came from Korea with at least a dozen ships, but because Japanese ships could not anchor in Ulleungdo, it was difficult to determine how many people visited the island, despite seeing a lot of people on the island.

⟨**Supplemental Source 2**⟩ *An Order Prohibiting the Crossing of the Seas*

When Hakushu and Inshu were under Matsudaira Shintaro's administration, there were instances in which Yonako merchant Murakawa Ichibe and Oya Jinkichi crossed to Takeshima and engaged in fishing up to the present, but bear deeply in mind that no further crossings to Takeshima will be allowed in the future.

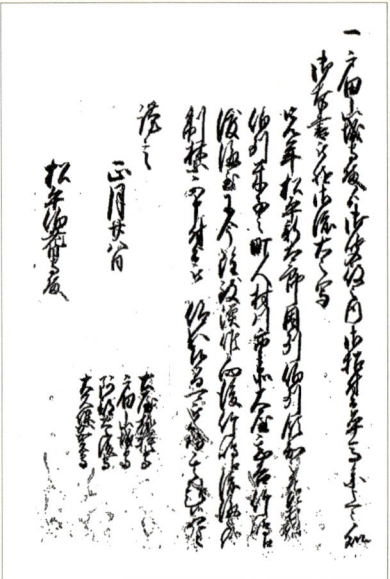

Source: Tottori Prefectural Museum

8

Jang Han-sang and *A Record of Incidents on Ulleungdo*

An Inspector of Ulleungdo Records Witnessing Dokdo

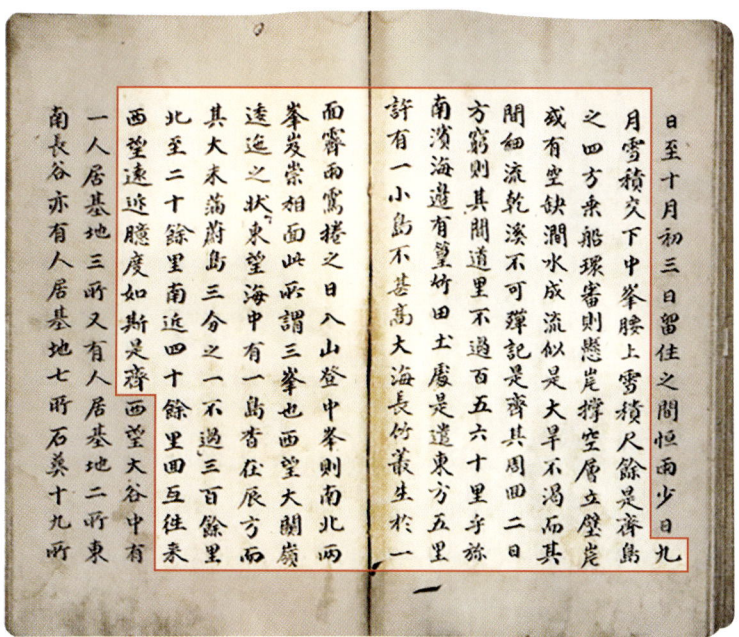

A Record of Incidents on Ulleungdo (Source: The Northeast Asian History Foundation)

September 1694: Snow piled up a few inches across the ridge of Middle Peak. As I traveled around the parameters of the island, I could see the jagged precipice supporting the thin air, and rocks were piled up like a wall and formed a hill. Water trickled between the rocks' empty spaces, and spouted out, refusing to dry up during a drought. There are also countless spouts of water and dry valleys which are too numerous to record. I was able to finish one tour of the entire island in two days, and the distance was less than 150-160 li. There was a small island about five li east, it was neither high nor big; there were long seaside bamboos thickly growing on one side. I entered the mountain and climbed up Middle Peak after the rain had stopped and clouds had cleared, whereupon I saw two peaks arise in the north and south, which collectively, along with Middle Peak are called "The Three Peaks." From afar, I could see Daekwan ridge curvaceously spread out to the West, and to the East, I could see an island in the middle of the sea, and it is distantly located in the Southeast, with a size that is less than a third of Ulleungdo, and was only about 300 li from Ulleungdo. The island stretches about 20 li north and almost 40 li south. I repeatedly visited the islands and measured distances throughout the island's parameter, and obtained such results.

A Record of Incidents on Ulleungdo is a document in which Jang Han-sang recorded his observations of Dokdo and its appearance. When the "Ulleungdo Controversy" broke out, the Korean government repented its negligence about Ulleungdo's affairs, and at the suggestion of Nam Ku-man, the Prime Minister, sent Jang Han-sang as Inspector.

Jang, who was serving as Samcheok's Provincial Magistrate, became appointed as Vice Admiral of Samcheok Port, and began inspecting Ulleungdo. He ordered the construction of new naval vessels to conduct inspection rounds, and collected provisions and supplies. He prepared one vessel to board his men, one cargo ship, four ships to serve as stations to distribute water, for a total of six ships along with 200 seok of food, Special Interpreter An Shin-hwi, Military Official Park Choong-jeong, assistants for shipmates, soldiers specializing in artillery and shipmates, for a total of 150 men.

After finishing his preparations for the inspection, Jang sent Military Official Choi Se-cheol of Samcheok in advance to explore navigational routes and patrol Ulleungdo to look for any Japanese citizens. On August 20, Choi led two light and fast fishing vessels to Ulleungdo, From Samcheok's Jang-ori-jin patrolled the area, and returned on September 1. On September 19, 1694 (Twentieth Year in the Reign of Sukjong), Jang departed from Samcheok and headed to Ulleungdo. After a severe ordeal of braving through strong torrential rain and high waves, he barely arrived at Ulleungdo. Jang stayed on the island for 13 days, inspecting every corner of the island. He left behind a report which meticulously recorded Ulleungdo's geography, climate, bamboo fields, remains of residential areas, ancient graves, ports which are suitable for docking, evidence of Japanese visits to the island, arboreal and animal species, and the quality of the soil.

What is especially significant is his record of what he witnessed all around while he was on top of Seongin Peak. He recorded that "5 Li to the East is a small island where seaside bamboos which are neither tall nor large, thickly and sporadically grow on one side of the island," and the island refers to what is now known as Jukdo (Bamboo Island). Bamboo trees still grow on the east side of the island. He also recorded that he "climbed up Middle Peak and saw Daekwan Ridge stretch out to the West." Jang then turned his attention to the Southeast and recorded the following about an island on the sea, which he could clearly see from afar. "When I climbed up Middle Peak and looked East, I could see an island located quite far to the Southeast. The island was less than a third of the size of Ulleungdo, and the distance between the two islands was a little more than 300 Li". This particular scene, which shows Jang Han-sang directly witnessing Dokdo from Ulleungdo and attempting to eyeball the distance to Dokdo and Dokdo's size, amazes anyone reading about it.

Having finished his inspections, Jang departed from Ulleungdo on October 4 and arrived at Samcheok Port the next day. He submitted a report and a map to the National Council, and in addition, presented three samples of sea lion fur, four bamboo shoots, and two pieces of sandalwood. Upon receiving Jang's investigative report, the Korean government decided that it was difficult to move people to Ulleungdo and construct a fort, and instead decided to hold inspection rounds every one or two years. The deployment of inspectors to Ulleungdo thereby became institutionalized.

Hanbandobawi(Provided by Ulleung County Office)

The Spread of Geographical Awareness About Ulleungdo and Dokdo

9

Frenchman D'Anville's
A Complete Map of the Korean Kingdom
18th century French Map Also Records Dokdo as Korean Territory

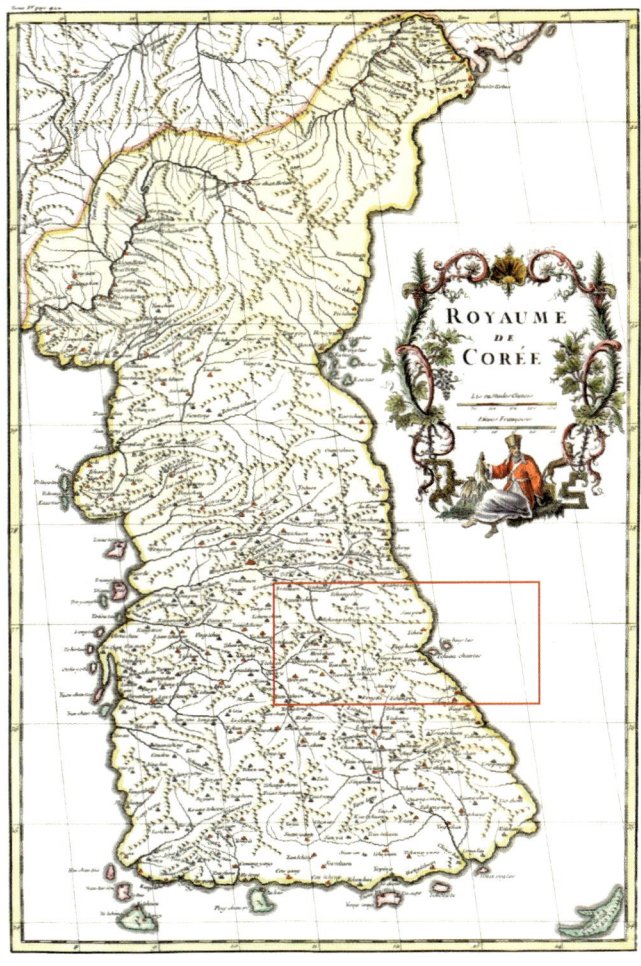

A Complete Map of the Korean Kingdom (44×60cm)
(Source: The Northeast Asian History Foundation)

Enlarged Portion of *A Complete Map of the Korean Kungdom* Showing Ulleungdo and Dokdo.

A Complete Map of the Korean Kingdom(1735) is the first map of Korea drawn by a European, and it notes Ulleungdo and Dokdo as Korean territory.

The practice of identifying Dokdo as Korean territory solidified as formal geographical knowledge and spread to Western cartographers. *A Complete Map of the Korean Kingdom* was created by noted French cartographer J. B. B. D' Anville based on the Qing Empire's *Complete Map of the Imperial Domain.** *A Complete Map of the Korean Kingdom* was first incorporated in Jesuit missionary Du Halde's *Record of China's Geography and History* (1735) and D' Anville's *New Collection of Maps of China* (1737). The map was the first among Western maps to record Korea as a separate realm, and hence is also the most meticulous in its portrayal of Korea. Dokdo is labeled "Tchianchantao" and Ulleungdo is labeled "Fanlingtao."

---

* Under the order of the Kangxi Emperor, French Jesuit priests, with assistance from some Chinese people, used the latest Western surveying techniques and completed a map of China's entire territory in a decade (1708-1718). D' Anville based his *New Collection of Maps of China* on this particular map, and all subsequent maps of China were based on this map.

From this evidence, we can infer that *Complete Map of the Imperial Domain*, on which D' Anville's *New Collection of Maps of China* was based, used Korean maps as its reference. Due to errors in comprehension or transcription, Usando was recorded as "Qianshandao," and Ulleungdo as "Fanlingdao," which are Chinese pronunciations of the Korean islands.* It is clear that contemporary Chinese as well as Western people who received information about the Orient from China, perceived Dokdo and Ulleungdo as Korean territory.

Moreover, *A Complete Map of the Korean Kingdom* provides valuable information about the Sino-Korean border. According to Supplemental Source 3, there are three different kinds of lines drawn on the northern border of the Korean Peninsula. The first line connects Baekdu Mountains with the Yalu and Tumen Rivers. The second line is a line of wooden fences stretching to the north from North Shineuiju, represented in black. Supplemental Source 4 shows the line extended as a series of wooden fences. This line represents the border of an area which the Qing sought to restrict the Han's access. The third line stretches north from the boundary between the Yalu and Tumen Rivers and is represented in orange, noting that the area south of the restricted territory belongs to Korea. Some Korean scholars refer to this orange-colored Sino-Korean borderline as the "Régis Line" after Father Jean-Baptiste Régis, who sent specific information on this area to France.

---

* An alternative theory suggests that it might have also been a case of misreading 鬱 as 攀.

⟨Supplemental Source 3⟩ *A Complete Map of the Korean Kingdom* : The Extension of the Régis Line

⟨Supplemental Source 4⟩ Wooden Fences Marking the Border of the Restricted Area ("Willow-Fence Border") in *A Complete Map of the Korean Kingdom*

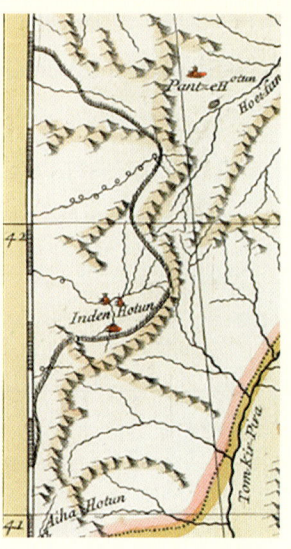

The Manchus, who chose Beijing as the capital of the Qing Empire, immigrated to China, which left their homeland of Manchuria as an empty patch of land. The Qing, wishing to prohibit the Han from entering the Manchus' homeland, established a restricted area and planted willow trees along the area's border and called it the "Willow-Fence Border."

10
The Grand Map of Korea
Dokdo is Properly Located East of Ulleungdo

The Grand Map of Korea (Source: National Museum of Korea)

Ulleungdo and Usando in *The Grand Map of Korea*

The Grand Map of Korea is the first map which properly represented the locations of Ulleungdo and Usando(Dokdo). An artist, under the auspices of the Office of Special Advisers, transcribed the map which was in the possession of Jeong Sang-ki's family after Jung's death in 1760, on a sheet of paper. Jeong is evaluated as having made a major breakthrough in the history of cartography in Later Joseon. Maps from the Early Joseon period tended to depict Usando(Dokdo) West of the Korean Peninsula, and Ulleungdo to the East, as in the case of *A Comprehensive Map of the Eight Provinces*, which is in *A New and an Extended Encyclopedia of Korean Geography*(1531). This practice was reversed in the Late Joseon Period, with Ulleungdo positioned West of the Korean Peninsula and Usando(Dokdo) positioned East of the peninsula, as seen in *The Grand Map of Korea*. Subsequent maps were drawn in the revised fashion, with maps such as *Haejwajeondo (A complete Map of Korea*) (Mid-19th century) being the most representative. (Confer Source 17).

The positional changes of Ulleungdo and Usando(Dokdo) can be closely associated with the Korean government's investigations of the islands' surrounding waters. During the Taejong and Sejong era, Ulleungdo residents were encouraged to relocate to mainland Korea, and consequently, Koreans' awareness and knowledge about the surrounding waters increased, leading to the creation of maps such as *A Comprehensive Map of Korea*, which depicts Ulleungdo and Usando(Dokdo) on the East Sea.

The perception acquired greater accuracy through the outbreak of the "Ulleungdo Controversy" (1693-1699), which arose when An Yong-bok and Park Eo-dun were kidnapped by Japanese fishermen, and subsequently evolved into a debate between Korea and Japan. The Korean government deployed Jang Han-sang, Provincial Magistrate of Samcheok, as Inspector to Ulleungdo to investigate the area, and Jang recorded witnessing a small island to the Southeast from Ulleungdo's Middle Peak(Seongin Peak). Moreover, An Yong-bok testified to Korean officials about Dokdo based on his experience of witnessing Dokdo or expulsing the Japanese from Dokdo. It was from such a process that maps such as *The Grand Map of Korea* came to be drawn.

Autumn(Provided by Ulleung County Office)

11

An Encyclopedia of Korean Customs and Culture

A Spread of Perceptions on Ulleungdo and Dokdo

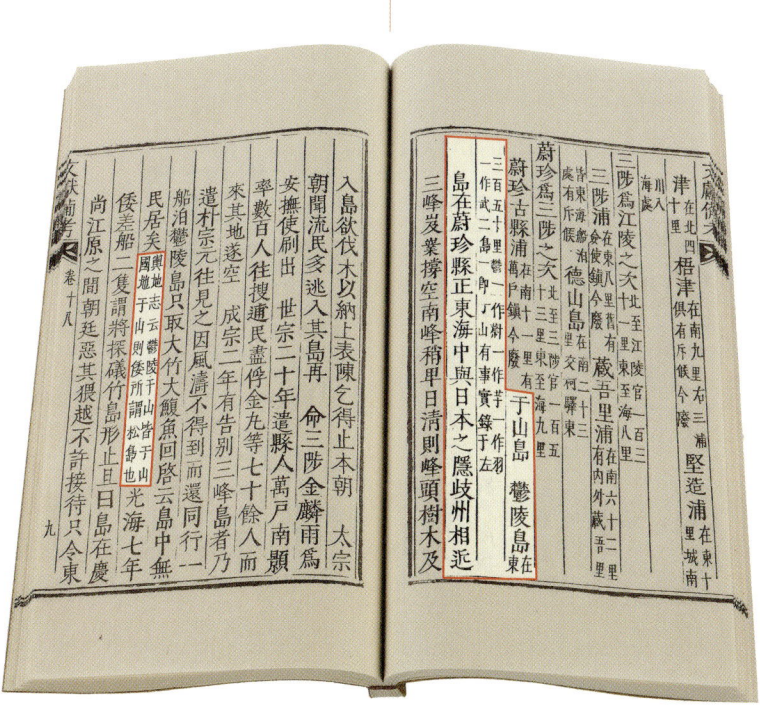

An Encyclopedia of Korean Customs and Culture (Source: Kyujanggak Institute for Korean Studies)

Usando and Ulleungdo

are two islands. One is called "Usan." The island lies in the middle of the sea directly east of Uljin County and is close to Oki Island. According to *An Encyclopedia of Korean Geography*, Ulleung and Usan are all part of Usan-Land's territory, and the Japanese call Usan "Matsushima."

An Encyclopedia of Korean Customs and Culture was published under the order of King Yeongjo as a book recording ancient and contemporary culture and customs of Korea. "Reflections on Korea's Territory," a chapter from the book, records geographical information on Korea's mountains, rivers, streams, and regions, along with the fact that Usando and Ulleungdo are Korean territories, as the translated excerpt above suggests.

The excerpt from above appears in the section on Uljin in "Reflections on Korea's Territory." Usando and Ulleungdo are entries, the main text is in a bold font, and the footnotes are in a small font. Shin Kyeong-jun, who authored *Reflections on the Geographical Boundaries of Korea*, took charge in publishing "Reflections on Korea's Territory," and thus, the translated excerpt is similar in content *to Reflections on the Geographical Boundaries of Korea*.

Shin Kyeong-jun left a detailed record of the "Ulleungdo Controversy," which was precipitated by the An Yong-bok incident, and like the excerpt from above, recorded Usando and Ulleungdo as Korean territories. Koreans became aware that the Japanese referred to Usando as "Matsushima" and as a result, Korean perceptions of Usando and Ulleungdo became highly clarified. *A General Introduction to Korea's Financial and Military Affairs*, published during the reign of Sunjo and *An Extended Encyclopedia of Korean Customs and Culture*, published during the era of the Korean Empire, inherited this perception.

12

Hayashi Shihei, *A General Cartographic Record of Japan* in *An Illustrated Description of Three Countries*

A Map from the Edo Period Also Records, "Dokdo is Korean territory"

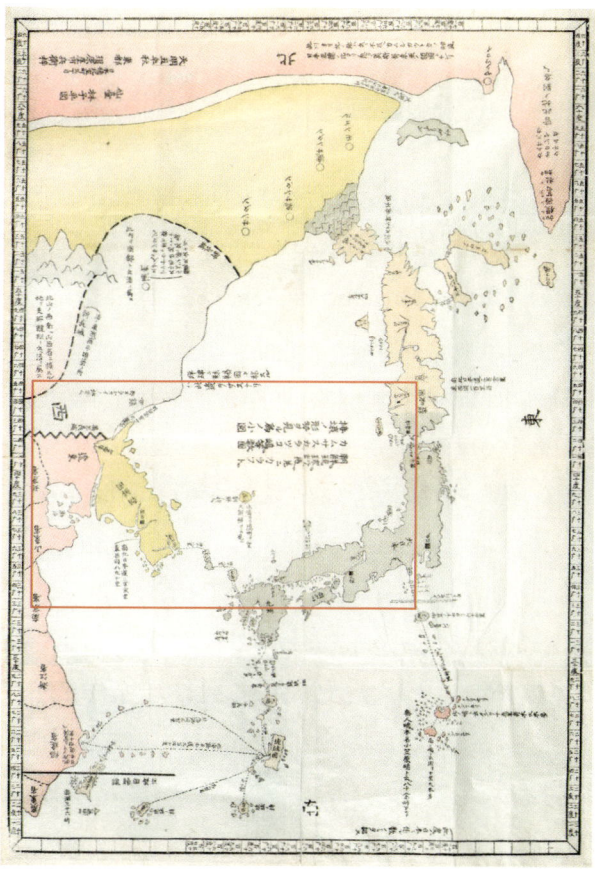

A General Cartographic Record of Japan in *An Illustrated Description of Three Countries*(50.5×72.5cm) (Source: The Northeast Asian History Foundation)

55

A General Cartographic Record of Japan in An Illustrated Description of Three Countries is an important primary source showing that Japan perceived Dokdo as Korean territory in the 18th century. The map, which Hayashi Shihei, a Japanese man, included in *An Illustrated Description of Three Countries*, is also known as *A Map of Three Countries Adjacent to Japan*. Hayashi, a *feudal retainer* from Sendai, wrote *An Illustrated Description of Three Countrie*s as a book after touring from Hokkaido to Kyushu. He described the customs and culture in three countries adjacent to Japan—Korea, Ryukyu, Ezo(Hokkaido), and Chishima's Karafuto, Etorofu, Urup, along with illustrations. This book records *A Complete Map of Ryukyu, An Illustration of an Uninhabited Island, A Complete Map of the Korean Nation*, and *A Complete Map of Ezo* in addition to *A General Cartographic Record of Japan in An Illustrated Description of Three Countries*.

A General Cartographic Record of Japan in An Illustrated Description of Three Countries has illustrations of a large island labeled "Takeshima" and a nameless small island, a note reading "these belong to Korea" to the left of both islands, and underneath the small island is a note which reads, "Oki Island and Korea are visible from this island". In addition, both islands, along with the Korean Peninsula, are colored in yellow.

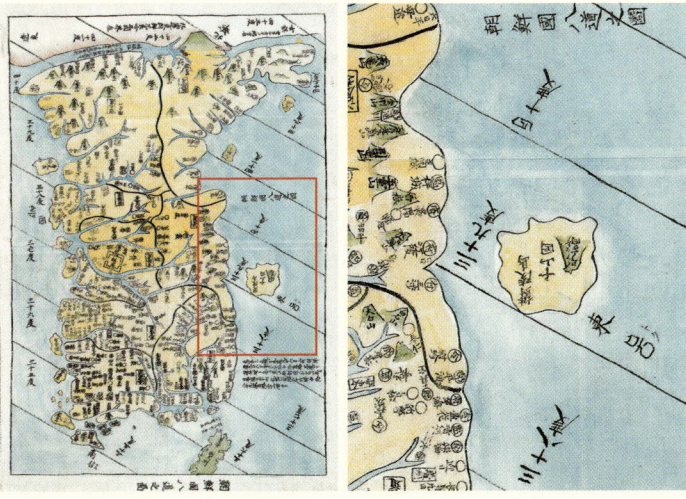

⟨**Supplemental Source 5**⟩ *A Complete Map of the Korean Nation* from *An Illustrated Description of Three Countries*

This map does not label Dokdo and labels Ulleungdo as 千山国 (an incorrect rendering of Usankuk 于山国) 弓嵩 is written next to Sung-in Peak, along with イソタケ in Hiragana. Whether "Isotake" is 弓嵩, which means "extremely high," read in Japanese transliteration (the Chinese characters are also read occasionally as "Isotake" in Japanese) or refers to "Isotakeshima" another Japanese name for Ulleungdo, is uncertain(54.2×77.2cm).

57

"Takeshima," illustrated as being on the East Sea, is Ulleungdo, and the nameless island is Dokdo. That the nameless island is Dokdo can be verified through a map called *A Grand Map of Three Countries— Hokkaido, Korea, and Japan* (1802), which is similar to *A General Cartographic Record of Japan in An Illustrated Description of Three Countries*, for like the latter, *A Grand Map of Three Countries* has illustrations of two islands on the East Sea, with the large island labeled "Takeshima" and the small island labeled "Matsushima." *A Grand Map of Three Countries* also has Oki Island and the Korean Peninsula adjacent to Takeshima on each side. To the left of Ulleungdo is a note which reads "Korea's" and the island is colored in yellow, the same color as the Korean Peninsula.

⟨**Supplemental Source 6**⟩ *A Grand Map of Three Countries—Hokkaido, Korea, and Japan*

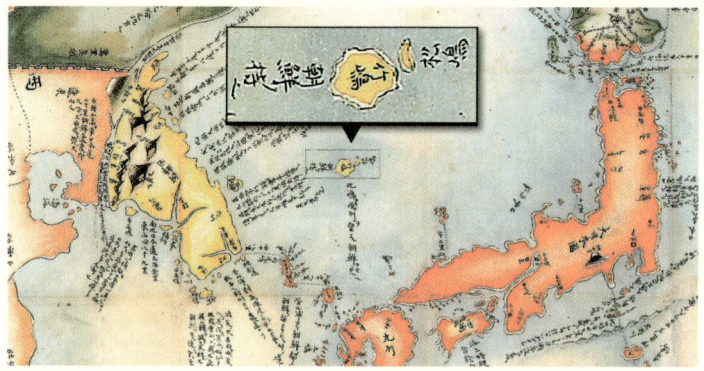

A map which Hayashi Shihei revised and complemented during his lifetime and published from 1801 to 1802 after his death(51.5×72.7cm).

13

A Comprehensive Map of Our Country
Ulleungdo and Dokdo on a Sea Labeled "East Sea"

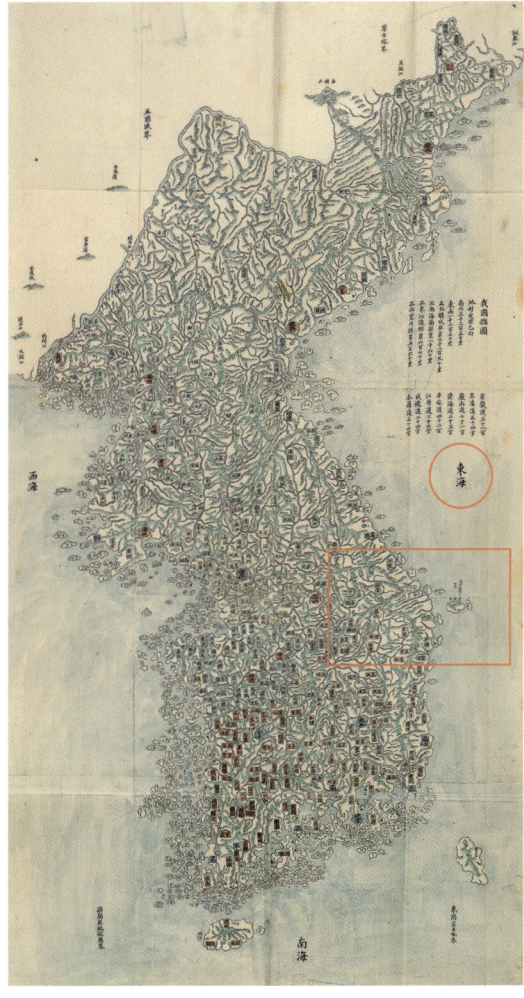

A Comprehensive Map of Our Country
(Source: Kyujanggak Institute for Korean Studies)

59

This map, through its illustration of Dokdo (Usando) to the East of Ulleungdo, is a source which demonstrates that the Korean government's geographical awareness became clearer after the outbreak of the An Yong-bok incident.

A Comprehensive Map of Our Country means that it is a map of the entire nation of Korea. The map, completed sometime between 1787 and 1799, is part of a three-page map called *A Map of Korea's Main Territories*. Although the cartographer is unknown, it is presumed that the map sought to replicate Jeong Sang-ki's *Map of Korea* or maps of a similar caliber at a reduced scale, with the accompanying information changed to suit the cartographer's objectives.

What distinguishes this map from other ancient maps is its use of more beautiful colors. In accordance with the Confucian Five Phases, the map records the names of counties in the East (Kangwon

Province) in blue, those in the West (Hwanghae Province) in white, those in the South (Jeolla and Kyeongsang Provinces) in red, northern ones (Hamkyeong Province) in black, and central counties (Kyeonggi and Chungcheong Provinces) in yellow—individual colors for the five directions. Mountains and streams are meticulously illustrated, and names of all islands, including small ones, are recorded. The seas are labeled as East Sea, West Sea, and South Sea according to their respective positions, and Dokdo is marked as Udo(于島). Usando, in reference to Jeong Sang-ki's *A Map of Korea*, is illustrated as being East of Ulleungdo, with a smaller size.

As such, this map shows the greater clarity of the Korean government's geographical knowledge after the An Yong-bok incident. In addition, the phrase "Could arrive in two days with the help of calm winds" shows that the Korean government had more specific geographical information, thanks to regular inspection rounds undertaken in Ulleungdo and Dokdo.

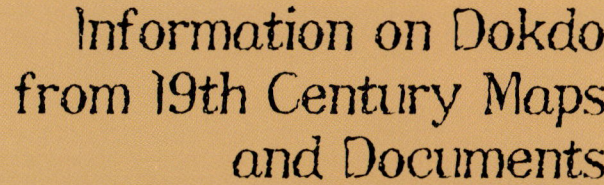

Information on Dokdo
from 19th Century Maps
and Documents

14

Kangwon Province in Haedongjeondo (A Complete Map of Korea)

Ulleungdo and Dokdo As Part of Kangwon Province

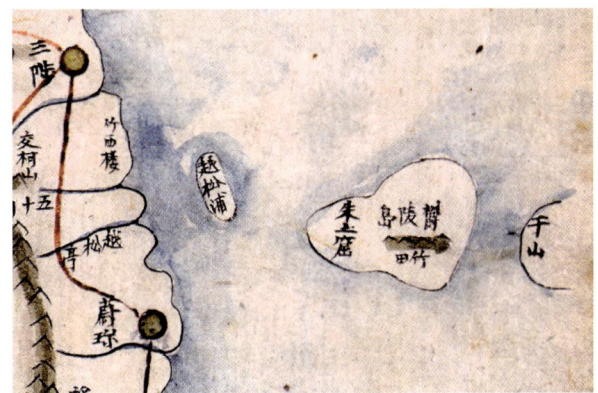

Haedongjeondo (*A Complete Map of Korea*)(47×30.5cm)
(Source: The Northeast Asian History Foundation)

The map *Kangwon Province* shows not only Ulleungdo and Dokdo as part of Kangwon Province but also emphasizes that fact by presenting enlarged illustrations of the two islands. This map is included in a colored edition of *Haedongjeondo* (*A Complete Map of Korea*). It is not an original work, and is instead presumed to have used the Korean government's maps as references and is a collection of maps on Korea's every county drawn as a series of paintings.

The cartographer who drew "Kangwon Province" is unknown, but was deeply influenced by Jeong Sang-ki's division of the Korean Peninsula into eight provinces, and like Jeong's *A Map of Korea*, features detailed representations of mountain ridges, streams, and road networks. Ulleungdo is drawn in the middle of the sea directly East from Uljin County, Kangwon Province, and Usan(于山, Dokdo) is drawn to the right of Ulleungdo, and thereby represented both islands to be part of Kangwon Province.

In addition, the map enlarged Ulleungdo and Usando and marked Ulleungdo's bamboo fields literally as "bamboo fields," and marked places where red dirt, which is used to make dyes, was available, as "Red Dirt Caves."

Since the mid-17th Century, the Korean government sent inspectors to manage territories. "Inspections" involved searching and punishing Korean residents or Wakou pirates who had secretly entered the island, a method which the Korean government regularly used to rule the island. Bamboos and red dirt were products which inspectors sent as tribute to the Korean government after inspecting Ulleungdo. The Korean government deployed the Provincial Magistrate of Samcheok or the Head Officer of Provincial Security in Wolseong and ordered them to perform inspections every two or three years. A notable feature of this map of Kangwon Province is that it marks Samcheok and Wolseong Ports (Wolseong is a port in Pyeonghae), which were points of departure for the inspectors, on the sea. This map serves as an excellent piece of evidence showing how Ulleungdo inspectors investigated Ulleungdo and Usando during the Late Joseon Period.

「Eclipse on a Cold Winter Night」(KimJungman, 2014)

15
Kim Dae-geon, *A Complete Map of Korea* (1845)
Map of Korea Presented to the West Shows "Usando"

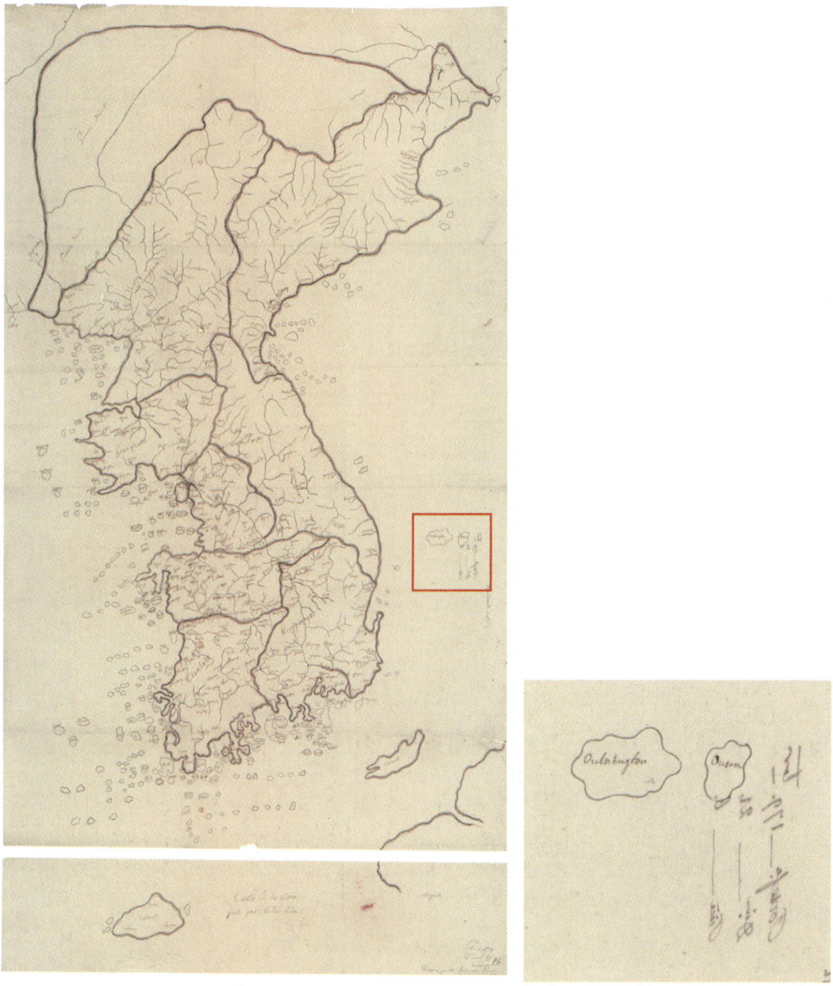

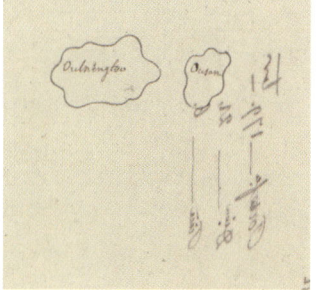

A Complete Map of Korea (Source: National Library of France)

A Complete Map of Korea not only marks Dokdo and Ulleungdo as Korean territories but is also the first map of Korea that was made in Korea and introduced to the West. Kim Dae-geon, the first Korean Catholic Priest, made the map in 1845, and the two copies, along with the original copy are stored in Paris's Bibliothèque nationale de France. Father Kim Dae-geon made this map to help French priests enter Korea and proselytize in the country. Hence, all place-names are written in English alphabets and there are 400 place-names in all. Of these, 260 are names of counties, 102 are names of islands, both of which comprise the majority. Bold lines on the map represent coastlines, national borders, and roads. A bold line in Manchuria seems to indicate the restricted area for the Han people.

According to one record, Kim Dae-geon copied a map from the Seoul Council. Judging from Kim's depiction of Korea's coastlines and northern border, Kim must have imitated a map similar in caliber to Jeong Sang-ki's *A Map of Korea*, which was widely in use during the 18th century.

This map is important in two ways. First, Usando, or Dokdo, is illustrated on the map, and its name is written in English alphabets to reflect the Korean pronunciation. Ulleungdo is rendered as "Oulangto," and Dokdo is rendered as "Ousan," or "Usan." Second, considering that most maps of East Asia produced in the West used Chinese or Japanese maps as primary references, Kim's map is the first map drawn by a Korean and introduced to the West from Korea. *A Complete Map of Korea* served as an important source of information on Korea's geography before the 1870s, when the West began to produce maritime charts on Korean waters. France and the United States, wishing to advance to Korea, produced imitated versions of this map to gather navigational information on the seas near to Korea.

16

Russia, *A Map of Korea's East Sea Coastlines*

Result of Modern Investigations of Navigational Routes Shows that Dokdo is Korean Territory

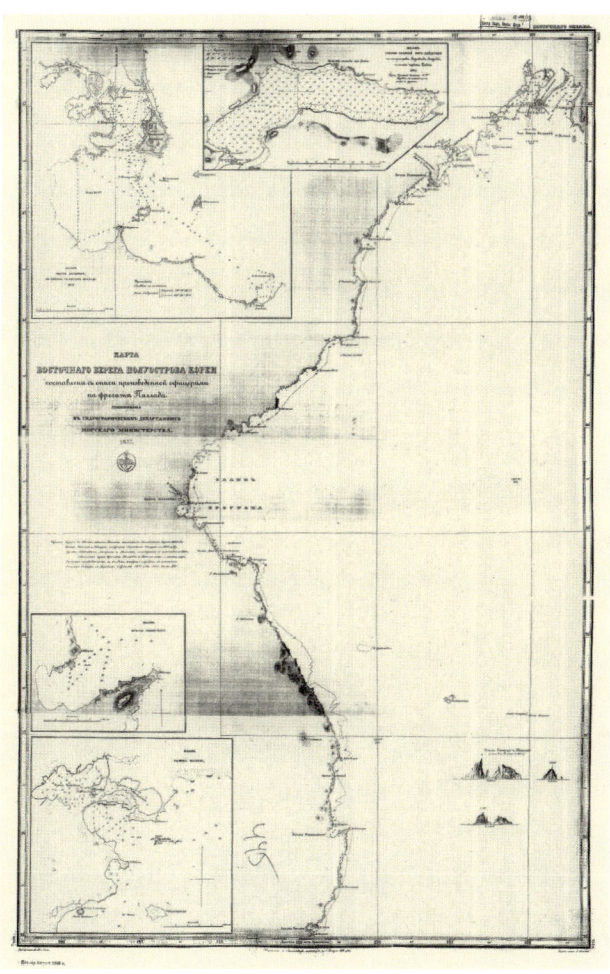

A Map of Korea's East Sea Coastlines (Source: Russian State Nabal Archives)

The Department for Waterways in the Russian Ministry of the Navy completed *A Map of Korea's East Sea* based on the Russian Navy's report of its navigation in the East Sea. This map shows that Russia officially recognized Dokdo as Korean territory.

In 1852, Nicholas I, aiming to "open" Japan, designated Admiral Putschachin as a Special Envoy. Admiral Putschachin assembled the battleship *Palada*, the *Corvette* (warship) *Olibutscha*, which was part of the Kamchatka Fleet, *The Vostok*, a sailing ship, and *The Duke Menshikov*, a transport ship, into a fleet and headed to Japan. The group of envoys accompanying Admiral Putschachin included Captain Unkovsky of the *Palada*, 22 officers, and 439 crew members. Among them were 6 officers and 37 marines aboard the *Vostok*. *Vostok*'s captain, Colonel Korsakov, was recommended by Putschachin himself. Father Abakkum of the Aleksandr Nevsky Monastery, was chosen to serve as the head interpreter of Chinese. The fleet sailed from Kronstadt (the naval base located on the sea near St. Petersburg) on October 7, 1852 and arrived at Nagasaki on August 10, 1853.

As negotiations with Japan became prolonged, the *Palada* decided to dock at Geomundo from April 2 to April 7, 1854. The *Olibutscha* discovered Dokdo while crossing the Korean Strait *en route* to the Tatar Strait (a strait between the Eurasian continent and Sakhalin Island) in the north, and named Dokdo's West Island, "Olibutscha," and East Island, "Menelai." The logbook records the following:

> Two rocks which we discovered in the morning were still visible for half a day, and everything about the rocks is now clear. The two acute

and naked rocks are about 300 sagens (642 meters) apart from each other. Among them, West Island is located at 37'13 Latitude North, 131'55 Longitude East. West Island, the taller of the two islands, was called *Olibutscha*, and East Island was called Menelai. Menelai was a name given to *Olibutscha* when *Olibutscha* was part of the Black Sea Fleet until 1846. Menelai is a reef which has risen above the water, and is located about 2 miles northwest from the *Olibutscha*. We discovered Olibutscha and Menelai Islands on a clear day from a distance of 30 miles. On April 6, 1854 (April 18 based on the solar calendar), our fleet spent about half a day on the sea, 4 miles west from Olibutscha Island.

The *Palada* Fleet was the first Western navy to separately ascribe names for Dokdo's West and East Islands.

In April, 1854, Admiral Putschachin, worried about the possibility of encountering icebergs and fog if his fleet crossed the Tatar Strait, decided to explore Korea's East Sea instead. From April 20 to May 11, 1854, the fleet investigated about 600 miles of coastal shorelines, from 35'30 Latitude North to 42'30 Latitude North and 131'10 Longitude East. In other words, the fleet investigated and took photographs of the Southeast of Korea's East Sea coast and parts of the East Sea's coastal areas located a few dozen miles further from the Manchurian coastlines. The *Palada* Fleet surveyed areas such as Songjeon Bay, located north from Yeongheung Bay in Yeongheung County, South Hamkyeong Province, and on clear days, investigated the underwater topography using diving equipment. The results were published in the January 1855 issue of the *Russian Naval Journal*:

> The *Vostok* observed Dagelet at 37'-22 Latitude North, 130'-56 Longitude East, and the island is circular in shape, with a parameter of about 20 miles. The coastline is narrow and was inaccessible. The highest peak on Dagelet measures about 2100 Feet (640.08 meters). The existence of Argonaut Island is dubious at best. The island was unobservable.
>
> The islands are covered with seabirds' excrement, and the *Olibutscha* discovered these islands at 37'14 Latitude North, 131'-57 Longitude East, and called them "Menelai" and "Olibutscha." The discovery of these islands is immensely helpful for navigations. These islands are far from adjacent islands and are positioned at crossroads for any ship traveling to the north of East Sea.

The Ministry of the Russian Navy's Waterways Department subsequently published *A Map of the Coastlines of Korea's East Sea* and officially recognized Dokdo as Korean territory. The 1857, 1868, and 1882 editions of the Russian Waterways Department's *A Map of Korea's East Sea Coastlines* are currently well-known. In the 1857 edition, which is currently being discussed, it is recorded that The *Olibutscha* discovered Menelai (East Island) and Olibutscha (West Island). However, in 1660, Lieutenant Colonel Sergeiev, Commander of the Russian Navy's Navigational Troops surveyed the topography of Dokdo. The 1868 edition introduced Sergeiev's survey in a map, and included various images of Sergeiev's surveys, which depicted Dokdo as seen from multiple directions. (Confer Source 16) The 1868 edition is the original copy of *A Map of Korea's East Sea Coastlines*, which Japan translated and published in 1876.

The 1882 edition was based on Russia's investigations from 1861 to 1880 and treats the *Olibutscha*'s discovery of Dokdo and Sergeiev's survey of Dokdo as mutually identical. This process illustrates how the discovery of Dokdo by the *Olibutscha* of the *Palada* Fleet, through the efforts of the Russian Ministry of the Navy's Waterways Department, became solidified as part of the Russian Government's official map. The Russian Government, through its continuous publication of such maps, confirmed that Dokdo is Korean territory.

「Here Come Spring Winds」(KimJungman, 2014)

17

Haejwajeondo (A Complete Map of Korea)

A Woodblock Map Which Marks Ulleungdo and
Dokdo Together and Provides Explanations About Them

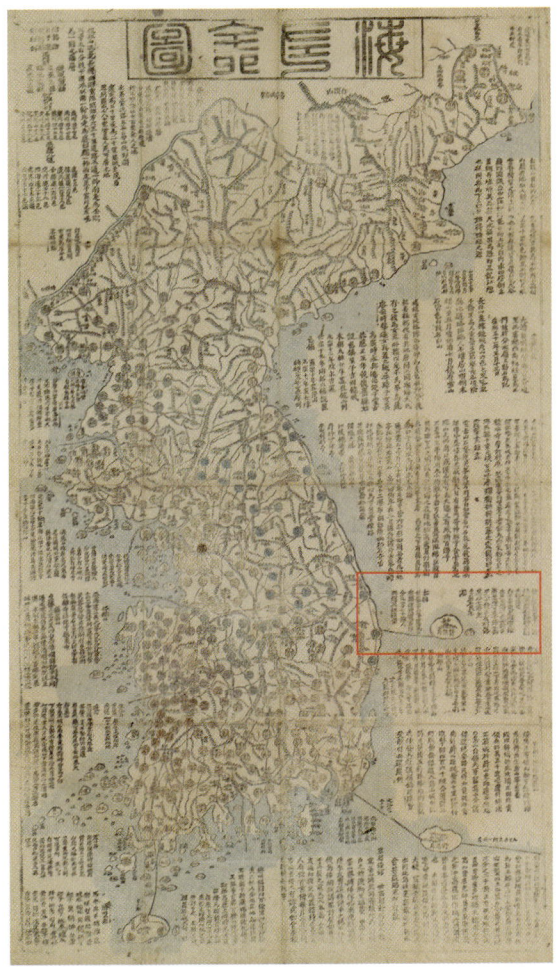

Haejwajeondo (*A Complete Map of Korea*)(55×98cm)
(Source: The National Folk Museum of Korea)

77

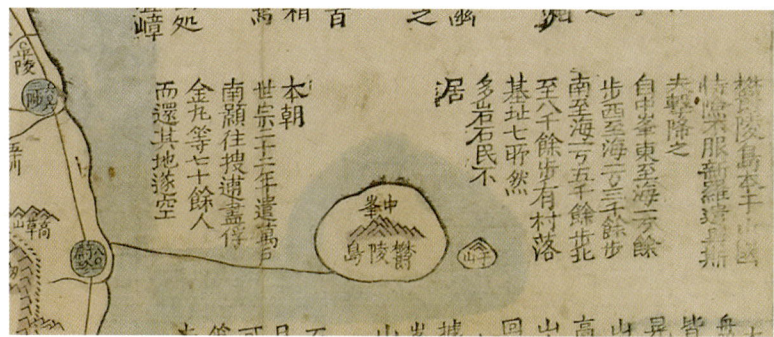

Magnified Portion of *Haejwajeondo* (*A Complete Map of Korea*)

Haejwajeondo (*A Complete Map of Korea*) is believed to have been made in the mid-19th century, and its creator is unknown. It portrays Korea's entire territory, and is especially noted for drawing Dokdo to the right of Ulleungdo and thereby clearly declares Dokdo to be Korean territory.

The title "A Complete Map of Korea" is written in seal characters above the map. The Chinese characters for Eastern Seas as used in the title literally means "to the left of the sea" and refers to Korea, since from a Sino-centric perspective of the world, if a person faces south and looks at the world, the Korean Peninsula exists across the sea from China (The Bay of Balhae and the Yellow Sea). Maps from the Joseon Dynasty used terms such as "East of the Sea" or "Eastern Nation" as alternative names for Korea.

Since this map is a wood block map, the map's precision is of a lesser degree compared with other contemporary maps, but important details about Korea's ecology, geography, and culture are impeccably recorded. Since wood block maps allowed for an easy reproduction of multiple copies, and because many copies of the map are still available today, the map was probably widely in use when it was first made.

This map depicts the Korean mainland as well as associated islands, and is especially faithful to Jeong Sang-ki's *Grand Map of Korea* from the 18th century in illustrating the contours of the Korean Peninsula. The peninsula's ecological elements, such as mountains, mountain ridges, rivers, and islands are depicted. Geographical and cultural elements expressed on the map include counties of each region, stations, military camps, forts, road networks, and other administrative and transportation matters, as well as military bases, thereby presenting a holistic image of the Korean Peninsula.

The empty spaces on the map are filled with information about Korea's history, the administrative areas of each dynasty, Mt. Baekdu, Mt. Myohyang, Mt. Keumkang, Mt. Songni, Mt. Jiri and other 20 or so famous mountains across the peninsula, as well as Jejudo, Ulleungdo, Heuksando, Deokjeokdo, and other islands, famous temples of Kangwon Province, the Eight Scenic View Points of the Eastern Provinces, such as Chongseokjeong, Kyeongpodae, Jukseoroo, and information about famous tourist attractions. There are also notes on territorial issues, such as the Baekdu Border Monument, which confirmed the Sino-Korean border, as well as brief histories of Tsushima Island, and Choryang's Japan House. It is particularly noteworthy that right next to the space in which Ulleungdo and Dokdo were drawn, the following words are written about Ulleungdo:

Ulleungdo was formerly known as Usankuk. When the island's residents refused to surrender due to their trust in the island's rough terrain, Silla sent Lee Sa-bu to invade the island and force its surrender. It takes about 10,000 steps from Middle Peak (Seongin Peak) to the East Sea, 13,000 steps to the West Sea, 15,000 steps to the South Sea, and 8,000 steps to the North Sea. There are seven village settlements, but no one lives in any of them due to the rocky terrain.

During the 22nd Year of King Sejong's reign, the government sent Chief Provincial Administration Officer Nam Ho to capture Kim Hwan and 70 other people, whereupon the territory was clear of residents.

The detailed nature of the above account about Ulleungdo's history and geography suggests that the map-maker perceives Ulleungdo as an important Korean territory. The map's record of Dokdo as an administrative territory associated with Ulleungdo confirms that Koreans have long regarded Dokdo as an associated administrative unit of Ulleungdo and as Korean territory.

18

An Intelligence Report on the Internal Affairs of Korea

Meiji Government of Japan Also Acknowledges, "Dokdo is Korean Territory."

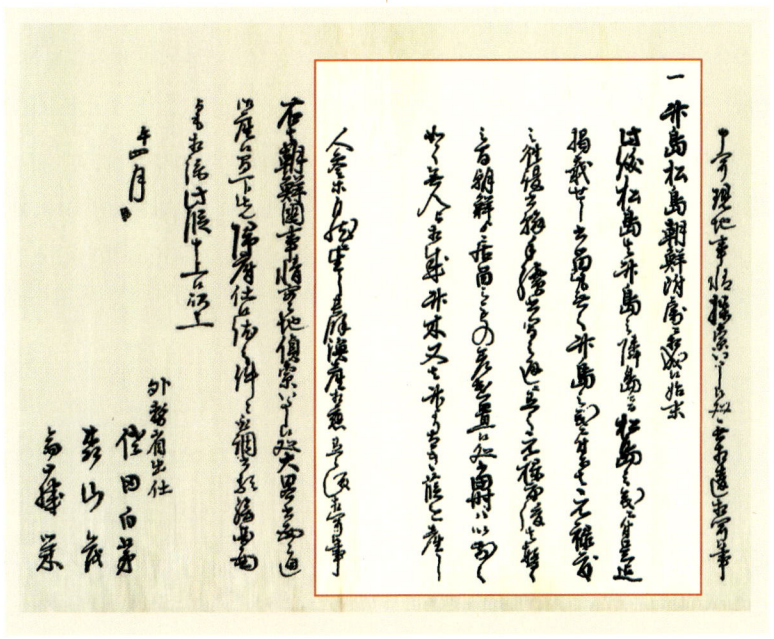

An Intelligence Report on the Internal Affairs of Korea (Source: National Archives of Japan)

How Takeshima(Ulleungdo) and Matsushima(Dokdo) Became Korean Territories

On this particular case, Matsushima is the neighboring island of Takeshima, and there are no written records on Matsushima. With regard to Takeshima, there are copies of official letters and procedural documents which the two countries exchanged during Genroku. The Korean government had sent people to inhabit these islands in the following years. However, no people inhabit the islands now, and there are rumors that the island has bamboo trees, reeds thicker than bamboo trees, and a copious amount of fish.

An Intelligence Report on the Internal Affairs of Korea(1870) (noted as *An Intelligence Report* hereafter) is a report which three officials from the Japanese Ministry of Foreign Affairs submitted to the Meiji government. This document informs that the Japanese Ministry of Foreign Affairs and the Dajokan, the highest institution in charge of public affairs in Japan both acknowledged Dokdo as Korean territory.

As Western imperialism arrived at the doorstep of East Asia, Japan became the first East Asian nation to open her doors to Western influence and transform into a modern nation-state. The Meiji government, upon its inception, attempted to apply modern international law in its relations with Korea and accordingly sent diplomatic documents aimed at establishing formal relations. However, the Korean government rejected Japan's offer because the documents contained inappropriate language. Japan's strategy of implementing *Seikanron** in the guise of establishing diplomatic relations with Korea had to undergo some adjustments after she failed to secure a treaty with Korea.

In order to examine past treaties between Korea and Japan and investigate Korea's internal conditions, the Japanese Foreign Ministry, under an order from the Dajokan, organized a 3-man investigation group comprised of Sada Hakubo, Moriyama Shigeru, Saito Sakae and sent them to Tsushima Island and the Japan House in Busan. They passed through Nagasaki and Tsushima and arrived at Busan on February 22, 1870, stayed at Choryang's Japan House, and carried out their investigation, and returned to Japan in March.

The three men investigated Korea on various aspects while staying at Busan and Tsuhima and submitted *An Intelligence Report* to the Japanese Foreign Ministry. This report is included in a book titled *Incidents in Korea* and is organized into 13 Subsections. In addition to diplomatic etiquette in Korea and conditions of Korea's trade, the report includes discussions of the Qing's influence over Korean sovereignty, conditions of Korea's military facilities and equipment, the existence of a suitable military base in Tsushima Island, and rumors that Korea was seeking Russia's protection, which suggests that such information must

---

* A theory from the 1850s which, supported by men such as Yoshida Shoin (吉田松陰), argued that Japan must invade Korea. The main architects of the Meiji Restoration put this theory into practice.

83

have been used to plan an invasion of Kang-hwa Island. The report was a pre-arranged investigative account of Korea's internal affairs which was to be used in planning an invasion of Korea.

Dokdo is mentioned in the last Subsection, "How Takeshima (Ulleungdo) and Matsushima(Dokdo) Became Korean Territory." Hence, under the assumption that Dokdo, along with Ulleungdo, is Korean territory, the Subsection describes the locations of the two islands and the situation in Ulleungdo. It was not originally part of the investigation and consists of only six lines and does not actually mention how Takeshima(Ulleungdo) and Matsushima(Dokdo) became Korean territories. It only mentions that "Matsushima is a neighboring island to Takeshima and there are no existing records about that island," and that with regard to Ulleungdo, "copies of official letters and procedural documents exchanged during Genroku era are available."*

It can be assumed that the investigation team must have meticulously studied the Tsushima Han's historical documents while investigating the diplomatic history of Korean-Japanese relations, along with the Korean-Japanese negotiations of the 1690s, or the Genroku era, which in turn, must have allowed them to record how Japan's border with Korea was determined. These assumptions seem valid; since they

---

* A Japanese imperial title ascribed to the period between 1688 and 1704; the "Ulleungdo Incident," which featured a clash between Korean and Japanese fishermen fishing near Ulleungdo, occurred from 1693-1699.

judged that Ulleungdo became Korea's territory after conducting a meticulous investigation, they were probably also able to determine that Matsushima(Dokdo), despite the absence of any written records, was also Korean territory.

That the investigation team carefully investigated the "Ulleungdo Controversy (Takeshima Incident)"(Confer Source 7) of the 17th century can be inferred by looking at "An Incident in Takeshima" in *An Investigative Report on Tsushima Island's Relations with Korea*, a special report attached to *An Intelligence Report*. This document contains six letters exchanged between the Korean government and the Tsushima Han, and thoroughly describes the circumstances behind Ulleungdo's categorization as Korean territory. The document shows that the Tsushima Han understood that Dokdo was an island adjacent to Ulleungdo and that crossing to Dokdo, along with Ulleungdo, was forbidden during the 1690s.

The investigation team understood the Tsushima Han's perception and regarded "the island next to Ulleungdo" and the Han's simple record about Dokdo more important than the existence or absence of documents about Dokdo, which led them to conclude that Dokdo is Korean territory. The absence of any documentation about Dokdo did not affect their strong perception that Dokdo was "an associated island of Ulleungdo" and that the two islands formed a pairing.

As such, *An Investigative Report on Tsushima Island's Relations with Korea* confirmed that Ulleungdo became Korean territory following negotiations after the outbreak of the "Ulleungdo Incident."

In addition, the investigation team concluded in its "Intelligence Report" that the "absence of records on Dokdo" and the fact that "Dokdo is a neighboring island of Ulleungdo" are sufficient reasons to regard Dokdo as Korean territory. In other words, "the Japanese Bakufu never claimed sovereignty over Dokdo" and "Dokdo is geographically close to Korea" became reasons with which the investigation team judged Dokdo as Korean territory.

Since *An Investigative Report on Tsushima Island's Relations with Korea* was submitted alongside *An Intelligence Report* to the Dajokan, the Meiji government must have also undoubtedly considered Dokdo as Korean territory. In short, the Meiji government inherited the Edo Bakufu's perception of Ulleungdo and Dokdo throughout the 1690s, before and after the "Ulleungdo Incident."

19

Japanese Ministry of the Navy, *A Map of Korea's East Coast*

The Japanese Navy Also Accepted the Russian Navy's Findings

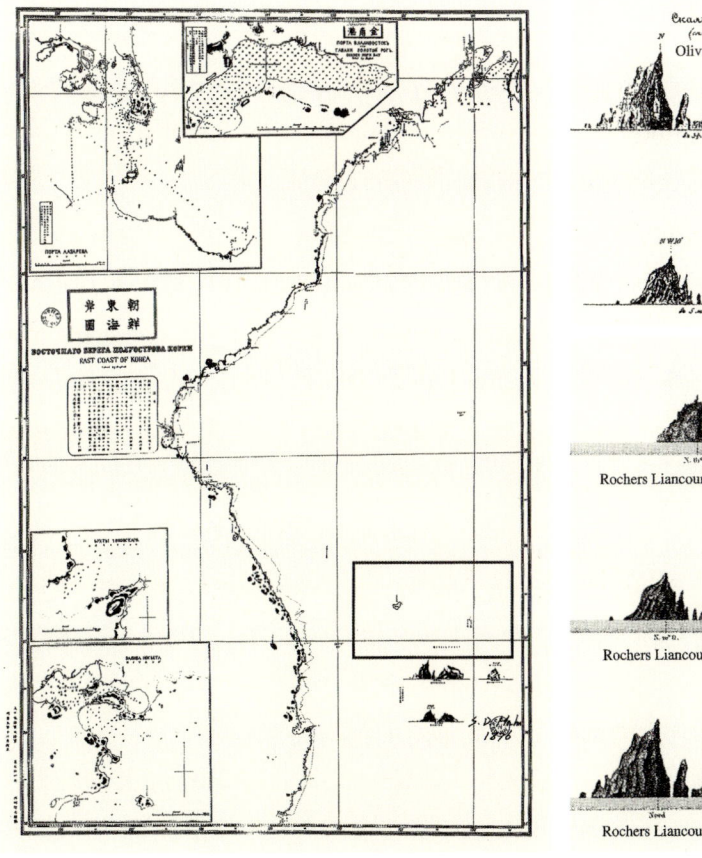

A Map of Korea's East Coast (63.5×98cm) (Source: The National Diet Library)

A Map of Korea's East Coast is a map which the Waterways Department of the Japanese Ministry of the Navy copied from the 1868 edition of the aforementioned Russian Navy's *A Map of Korea's East Sea Coastlines* and translated and published in Japanese. (Confer Source 16) This fact implies that Japan accepted Russia's geographical perception that Dokdo is Korean territory.

Shortly after the Meiji Restoration, the Waterways Department of the Japanese Ministry of the Navy made maritime charts by translating and editing various maritime charts which were surveyed and investigated by British, French, Russian, American and other Western naval vessels. The Japanese Waterways Department especially used Russian maps as references for regions surrounding the East Sea.

In addition to the East Sea coastline stretching from Vladivostok to Ulsan Harbor, Ulleungdo and Dokdo are also marked on the Japanese maritime chart. With regard to Dokdo, the chart not only has a map of Dokdo but also included an illustration of the island's contours. Lieutenant Sergeev of the Russian Navy left this illustration after surveying Dokdo's terrain in 1860. The illustration contains meticulous drawings of Dokdo's East Island, known as "Olibutscha" and West Island, known as "Menelai" to the Russians. In addition, the illustration includes images of Dokdo seen from a ship 14 miles, 5 miles, and 3.5 miles away respectively, enabling sailors to use these images as references during actual navigations.

The inclusion of Dokdo and Ulleungdo in *A Map of Korea's East Sea* suggests that Russia perceived these islands as Korean territories, and Japan's translation of the Russian map suggests that Japan accepted Russia's perception. The Waterways Department of the Ministry of the Japanese Navy included the images of Dokdo in *A Map of Korea's East Coast* rather than *A Map of Japan's Northwest Coast* and marked Dokdo in the same manner as *A Map of Korea's East Sea* when the Department republished *A Map of Korea's East Coast* in 1887. Dokdo was marked as Korean territory in *A Gazetteer on Korea's Waterways* (1889) (Supplemental Source 7), and this practice was continued even after Japan annexed Dokdo in 1907.

⟨Supplemental Source 7⟩ *A Gazetteer on Korea's Waterways* (1899)

The second edition of *A Gazetteer on Korea's Waterways*, published in 1899, introduces Dokdo as the 'Liancourt Rocks,' a name given by the French ship Liancourt when it first discovered Dokdo and gave the ship's name to it, and that Russian and British ships gave their ships' names to Dokdo when they discovered it.

⟨**Supplemental Source 8**⟩ *A Gazetteer on Korea's Waterways* (1907)

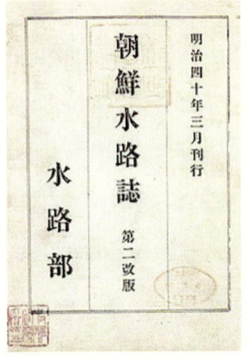 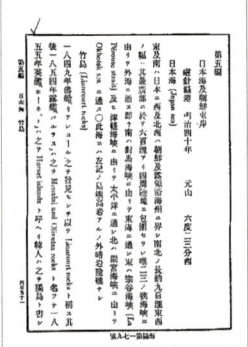

It records that French, Russian, and British navigational vessels respectively refer to Takeshima (Dokdo) as the Liancourt Rocks, Menelai and Olibutscha, and the Hornet Islands; Koreans call it Dokdo and the Japanese call it "Liacor Island." Despite the fact that two years had passed since Japan annexed Dokdo, the second edition of the Gazetteer still recognized Dokdo as a Korean territory(2nd edition published in the 40th Year of Meiji, 21×29.5cm, 1907).

20

Dajokan Directive and *A Map of Isotakeshima*

Japan's Top Administrative Institution, "Japan Has No Business With Dokdo"

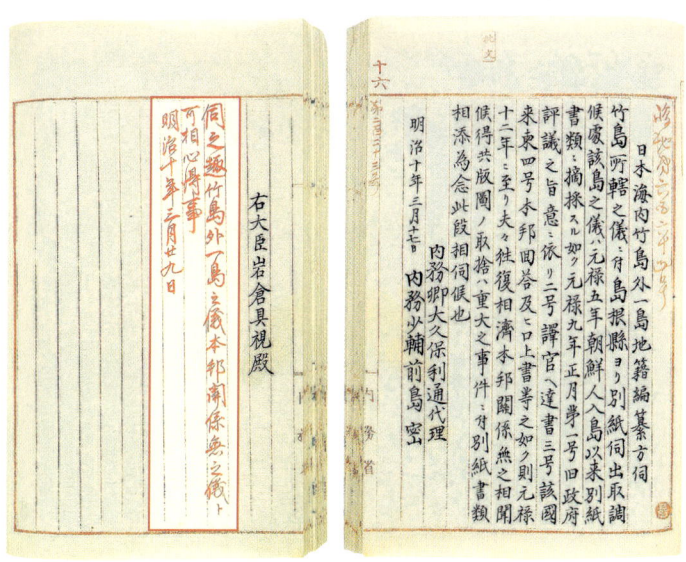

Dajokan Directive (Source: The Northeast Asian History Foundation)

Bear in mind that with regard to the inquiry about Takeshima (Ulleungdo) and another island, our country (Japan) has no business with these islands. March 29, Tenth Year of Meiji (1877)

The Korean nation is visible from IsoTakeshima(Ulleungdo) at a distance of 50 li (200 KM)*
Matsushima(Dokdo) is 40 li (160 KM) Northwest of Isotakeshima (Ulleungdo).
Matsushima is 80 li (320 KM) away from Togo of Oki Island.

To consolidate its authority, the Meiji government wished to produce an accurate cadastral map, which would serve as a national land registry. Therefore, the government ordered all prefectures to investigate and register their maps and plots of land.

However, registering Takeshima(Ulleungdo) proved to be a problem for Shimane Prefecture. On October 5, 1876, the Home Ministry sent a request and a statement of inquiry to organize the registration of Takeshima(Ulleungdo). Therefore Shimane Prefecture conducted an investigation into the Oya Family's records on their crossings to Ulleungdo and submitted a questionnaire which included *A Summary of Reasons and Causes* and *A Map of Isotakeshima* to the Home

---

* 1 li is still about 400 meters in Korea, but in Japan, it is about 3.9 KM. The actual distance between Ulleungdo and Jook-byeon is about 130.3 KM, the distance between Ulleungdo and Dokdo is about 87.4 KM, and the distance between Dokdo and Oki Island is about 157.5 KM, meaning that there is a significant difference in survey results from that period and the actual distances.

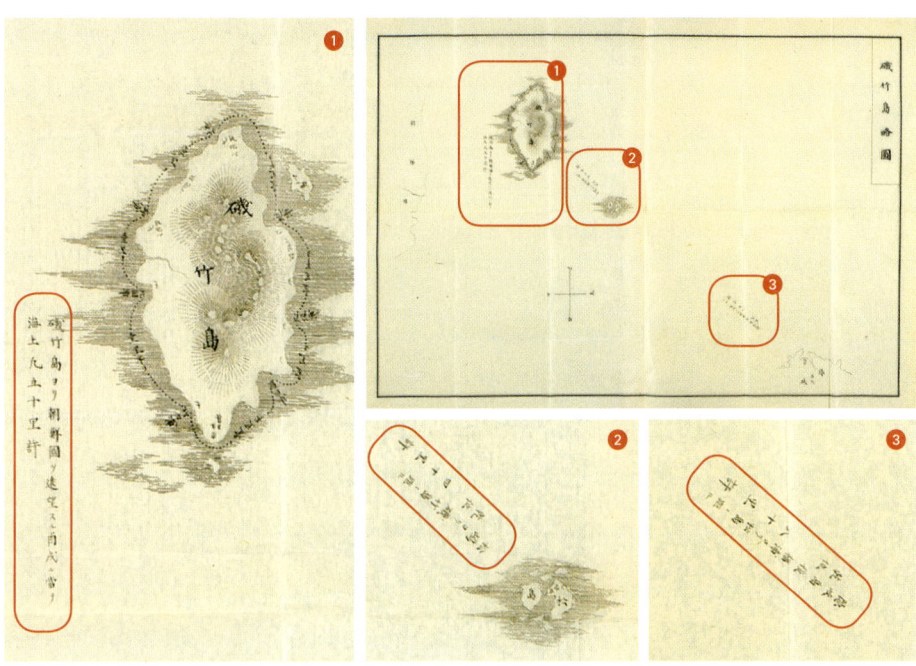

Map of Isotakeshima

Ministry. The questionnaire, with its long title of *An Inquiry About Methods on Publishing a Registry on Takeshima and Another Island in the Sea of Japan*, requested that the government must make a decision about whether Takeshima(Ulleungdo) and one other island(Dokdo) must be included in or excluded from Shimane Prefecture's land registry.

Shimane Prefecture's inclusion of the term "one other island," despite the absence of such a term in the Home Ministry's statement of inquiry was probably due to the Oya Family's documents mentioned "Matsushima adjacent to Takeshima" and "Matsushima within Takeshima," which led the prefecture to perceive Matsushima and Takeshima as constituting a group or Matsushima as a part of Takeshima. As *A Summary of Reasons and Causes* and *A Map of Isotakeshima*, attached documents to the Shimane Prefecture's questionnaire, clearly show, Shimane Prefecture considered "one other island" or Dokdo as a part of Ulleungdo.

A Summary of Reasons and Causes consists of "A Summary of Origins," which explains why Japanese citizens went to Dokdo, "A Permit to Cross the Seas," "Circumstances on Prohibiting the Crossing of the Seas," "An Order Prohibiting the Crossing of the Seas," and "Reflections," which is a statement emphasizing Shimane Prefecture's ties with Ulleungdo and expressing Shimane Prefecture's disappointment with the Bakufu's ill-considered policy of prohibiting the crossing of the sea between Shimane Prefecture and Ulleungdo.

After receiving Shimane Prefecture's questionnaire, the Home Ministry concluded after conducting an internal investigation for five months about Korea and Japan's exchange of letters during the Ulleungdo Incident, that Ulleungdo was not Japanese territory, that Ulleungdo and Dokdo were one island, and that they had no relation to Japan's interests. However, because possessing a territory was an important issue, the Home Ministry decided to make a request to the Dajokan to hand down the final decision.

The confidence with which the Home Ministry made its decision about the classification of Ulleungdo and Dokdo within such a short period of time rests on several causes. The Home Ministry utilized *Views on Isotakeshima* (1875), published by the Geography Department of the Home

Ministry, in conducting the Home Ministry's investigation and writing the Home Ministry's request for the Dajokan's review. Even a year before its request to the Shimane Prefecture and Shimane Prefecture's submission of its documents, the Home Ministry had collected documents related to the Bakufu, the Tsushima Han government, the Tottori Han government, and other ancient historical documents to publish *Views on Isotakeshima*. The ancient historical documents in this collection perceive Takeshima and Matsushima as a pair of islands, and clarify that these islands are not Japanese territories. Therefore, by this time, the Home Ministry was already well aware that Takeshima and Matsushima are not Japanese territories.

Upon receiving the Home Ministry's inquiry, the Dajokan swiftly made a decision. Despite acknowledging that "acquisition of territories is a serious issue," the Dajokan concluded within three days in its directive that "Takeshima and one other island have no business with Japan." The Home Ministry and the Dajokan autonomously decided on Dokdo's classification and declared that it is a matter unrelated to Japan. The problem of classifying Dokdo and Ulleungdo was thereby completely solved.

Despite these facts, some Japanese scholars still insist that "one other island" does not refer to Dokdo. Yet, if one sees *A Map of Isotakeshima* attached to the statement of inquiry, which shimane Prefecture submitted to the Home Ministry, everything is quite clear. Matsushima as it appears on the map consists of two islands to the East and West, which must be Dokdo. Japan's Home Ministry completed this map based on the Oya Family's drawings, which the Home Ministry submitted to the Dajokan. Hence, it can be assumed that the Home Ministry and the Dajokan shared the same perception.

Winter(Provided by Ulleung County Office)

21

The Japanese Army, *A Complete Map of Great Japan*

Dokdo Is Not Drawn on an Early Modern Japanese Map

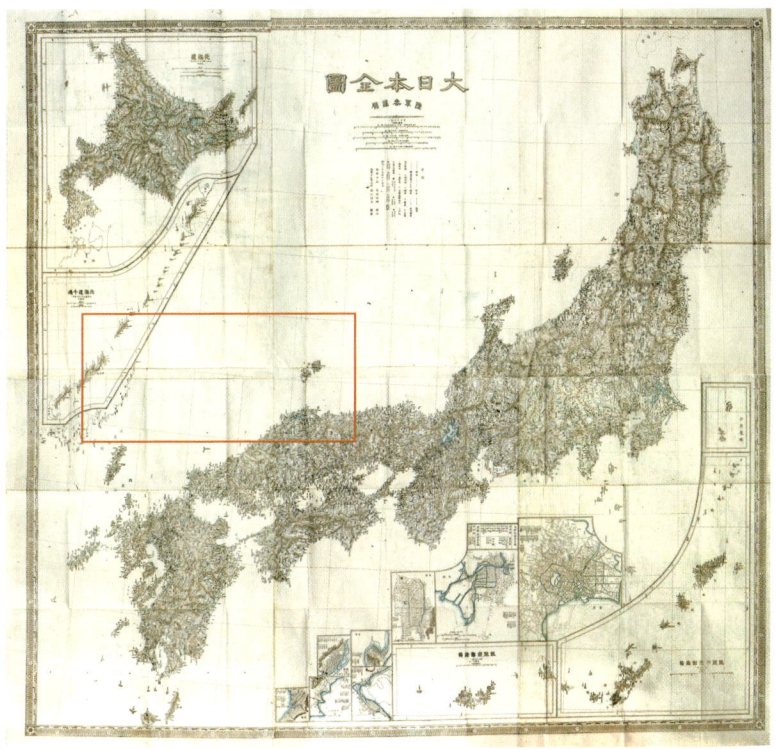

A Complete Map of Great Japan (123×115cm) (Source: Kyoto University)

A Complete Map of Great Japan demonstrates that Dokdo, in addition to Ulleungdo, was excluded from Japan's territory in maps created by the Japanese government. This fact remains valid despite a French cartographer's creation of *A Complete Map of Great Japan* based on an official map from the Tokugawa period. This map is a modern Japanese map created by Kimura Nobuaki, Director of the Land Survey Department of the Japanese Army Headquarters, and Shibue Nobuo, an employee, at a ratio of 1:1600,000.* Kimura was in charge of French cartographic methods in the Japanese Army. This map specifically illustrates the Japanese mainland, but does not mark

---

* Generally, maps of the Korean Peninsula drawn on A4 paper are at a scale of 1:4,000,000.

Ulleungdo and Dokdo.

A Complete Map of Great Japan was based on a map created by Ino Tadataka, the cartographer in charge of making official maps during the Edo Period. Ino Tadataka who retired from business at age 50 became interested in astronomy, and then turned his attention to cartography. He was able to complete Japan's first scientifically surveyed map after traveling around Japan on ten occasions during a period of 17 years, and his students edited and completed *A Grand Map of Japan's Coasts and Main Territories*. The map has but an error of 1/1000 compared with measurements produced using modern surveying techniques. The map was rarely used during the Edo Period, but the new Meiji government published it and used it as a fundamental map for military, educational, and administrative purposes. Ino Tadataka remains one of the most revered figures in Japan today.

Other maps created through the Bakufu's cartographic projects, or official maps, did not depict Ulleungdo or Dokdo.

Neither Edo Bakufu's Maps which were created through the use of actual measurements, nor the Meiji Government's maps which reflected European cartographic techniques perceive Dokdo as Japanese territory.

Japan's Intrusion and the Korean Empire's Resolute Management of Territories

22

Lee Kyu-won, *An Investigative Diary on Ulleungdo*

The Korean Government Actively Responds to Japan's Intrusion on Ulleungdo

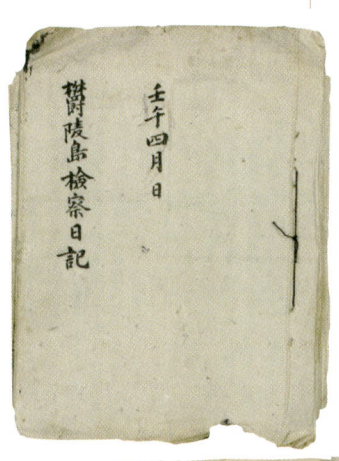

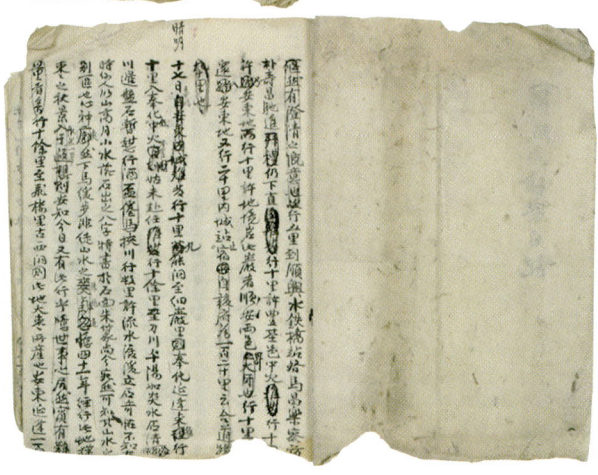

An Investigative Diary on Ulleungdo (Source: Jeju National Museum)

An Investigative Diary on Ulleungdo is a diary which records every part of a round-trip which included departing from Seoul for Ulleungdo, investigating Ulleungdo and returning to Seoul and served as an important *in situ* record which became the foundation for Kojong's decision to strengthen his administrative authority.

Around 1880, the Japanese were illegally entering Ulleungdo and engaging in logging. An Inspector in charge of Ulleungdo had also reported that the Japanese were making unauthorized entries into Ulleungdo and cutting down trees to make boats, and this became a diplomatic issue between Korea and Japan. In May 1881, Kojong named Lee Kyu-won as Investigator for Ulleungdo and instructed Lee to assess the situation in Ulleungdo.

Lee met Kojong on April 7, 1882 and departed from Seoul on April 10, 1882. On April 29, 1882, Lee set sail for Ulleungdo from Koo-san Port(Uljin) and investigated Ulleungdo until May 12. Lee arrived at Little Yellow Mud Hill (currently Hakpo, Ulleung County) and inscribed on a rock, "Lee Kyu-won, Investigator for Ulleungdo, May, 1882." (See Supplemental Source 9)

Lee made a draft of a report in May of 1882, and wrote an *Investigative Diary on Ulleungdo* (1900). Lee left illustrations depicting the roughness and smoothness of the terrain, fertile and infertile soil, and painstakingly identified potential residential areas, the island's produce, and species of birds. Lee selected an area around Ulleungdo's Nari Basin and recommended it for developing land and managing farms. Kojong issued an order to develop farmlands based on Lee's report and an interview with him and encouraged relocation to Ulleungdo from 1883 in incremental stages. The Korean government made efforts to expulse Japanese citizens who were illegally engaged in logging at Ulleungdo, established a new administrative system for Ulleungdo, and gathered Korean citizens wishing to relocate to Ulleungdo. The Korean Empire, which began developing Ulleungdo since September 1899, implemented an *in situ* investigation of Ulleungdo by assigning an Overseer as an investigational associate. As a result, the Korean Government was able to institute Imperial Decree No. 41 of the Korean Empire towards Dokdo and Ulleungdo on October 25, 1900.

⟨Supplemental Source 9⟩ Inscription Left on Rocks at Hakpo, Tae-ha Village, Ulleungdo

Ko Jong-pal(高宗八), former Keeper of the Gates, who participated as a member of Lee Kyu-won's(李奎遠) investigation team, also inscribed his name alongside that of Lee.

23

Imperial Decree No. 41 of the Korean Empire

The Korean Empire Declares Dokdo as Its Territory

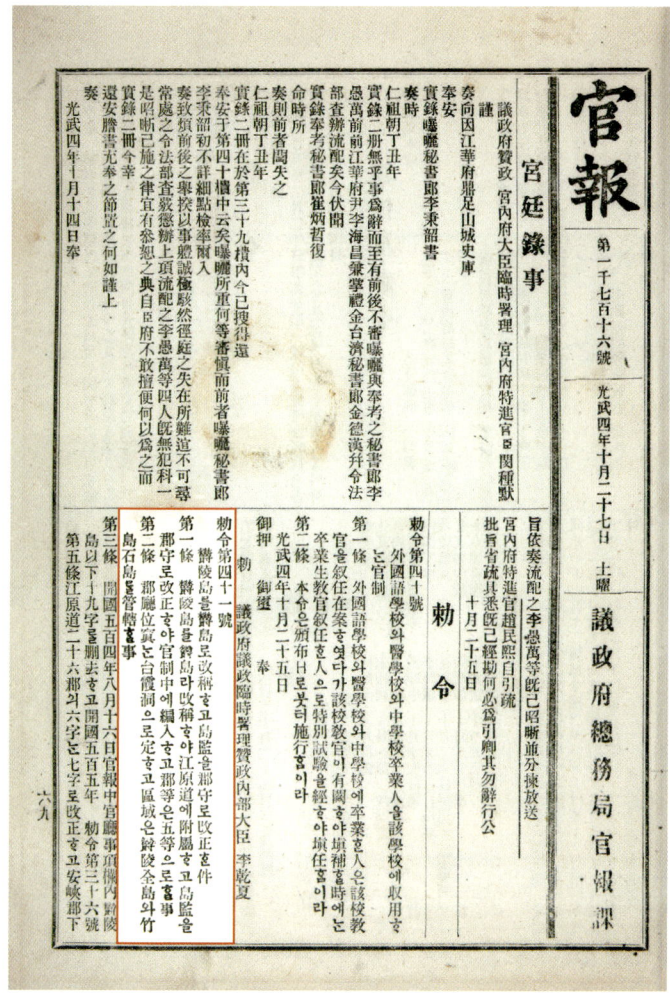

Imperial Decree No. 41 of the Korean Empire (Source: Kyujanggak Institute for Korean Studies)

Imperial Decree No. 41

Ulleungdo shall be renamed "Uldo," and the rank of Island Governor shall be elevated to County Mayor

Article 1: Ulleungdo shall be known as Uldo and become part of Kangwon Province. "Island Governor" will be elevated to County Mayor and be designated as a formally ranked official. The county will be classified as a Level 5 county.

Article 2: The County Office shall be located in Tae-ha Dong and the entire island of Ulleungdo, Jukdo (Bamboo Island), and Seokdo (Stone Island) will be the Office's administrative region.

October 25, Fourth Year of Kwangmu (1900)

Imperial Decree No. 41 of the Korean Empire, announced in an official gazette, shows that the Korean Empire fully expressed territorial sovereignty over Ulleungdo and Dokdo as a modern legal decree. The Korean Empire prepared an *in situ* investigation of Ulleungdo during a discussion about Ulleungdo's administrative reforms. Illegal logging activities of Japanese residents in Ulleungdo was a diplomatic problem between Korea and Japan at the time. Under such a circumstance, the Korean government decided to deploy an Overseer to Ulleungdo. Associate Overseer Woo Yong-jeong investigated and submitted a report on illegal entries of Japanese citizens to the island from June 1 to June 5, 1900.

Woo Yong-jeong, Associate Overseer of Ulleungdo in the Overseer's Office of the Korean Home Ministry, Kim Myeon-soo, Chief

Superintendent of Dongrae E. Laporte, Secretary of the Busan Maritime Customs Office, Akasuka Shosuke of the Japanese Consulate in Busan and several others boarded a ship named *Chang-ryong* and headed for Ulleungdo. On May 31, 1900, Woo arrived at Do-dong Port, where Ulleungdo's Governor Bae Kye-ju was residing.

Woo Yong-jeong recorded the following schedule for the investigation from June 1 to June 6. In the presence of Laporte, Woo investigated Japanese citizens with Akasuka. On June 4, 1900, Woo boarded the *Chang-ryong* and patrolled the entire island of Ulleungdo. Woo spent June 5 answering inquiries from Ulleungdo's residents. After completing his investigative report, Woo departed from Ulleungdo and returned to Busan at 10 AM on June 6, 1900.

Woo Yong-jeong and Kim Myeon-soo directly witnessed Japanese residents who were illegally residing on Ulleungdo. In June, 1900, their number was 144. It was revealed that they built 57 grass huts and possessed 11 ships. Kim Myeon-soo quickly requested the Japanese Consulate in Korea to encourage the Japanese residents to return to Japan.

Woo Yong-jeong recommended that the Governor of Ulleungdo reinforce people serving in low-ranking offices such as Director of Labor Relations, Secretaries, and Assistants, and suggested specific improvement methods. To manage the County Office of Ulleungdo, Woo suggested that it as well collect 2 heads of barley and beans from four hundred households, totaling to 80 Seok, or about 11,040 KG.

Woo Yeong-jong had independent authority over Ulleungdo such that he was able to personally decide the purchase of a ship called the *Kaewon*. He made considerable efforts to strengthen the Korean Empire's administrative authority over Ulleungdo and organize the island's infrastructure. He pointed out the fact that his request to the central government to reform Ulleungdo's administration was being suspended and personally suggested methods on how to prepare funds. (Confer Supplemental Source 10)

Woo Yong-jeong's deployment was a part of the Korean government's plan to reform Ulleungdo's administration. This suggests that the Korean government recognized the importance of Ulleungdo and deployed Woo to strengthen the government's administrative authority. As a result of such activities, the Korean Empire issued Imperial Decree No. 41 on October 25, 1900 and on October 27, published the decree in the official gazette.* The decree's title was "Ulleungdo shall be renamed 'Uldo,' and the rank of Island Governor shall be elevated to County Mayor." This decree defined the administrative area of Uldo as comprising the entire island of Ulleungdo, Jukdo, and Seokdo.

Jukdo referred to "Bamboo Island" next to Ulleungdo and Seokdo (石島) referred to "Stone Island" or Dokdo.

---

* A gazette is an official paper which announces issues which the government deems necessary to publicize to the people and gets edited and published as the government's official newspaper. In addition to constitutional reforms, various decrees, notices, the condition of the national treasury, treaties, investiture, dismissing of officials, congressional issues, and issues addressed in administrative offices and all legal decrees were announced through gazettes. The first modern gazette was founded on June 21, 1894.

⟨Supplemental Source 10⟩ A Request from the Minister of Internal Affairs (1900)

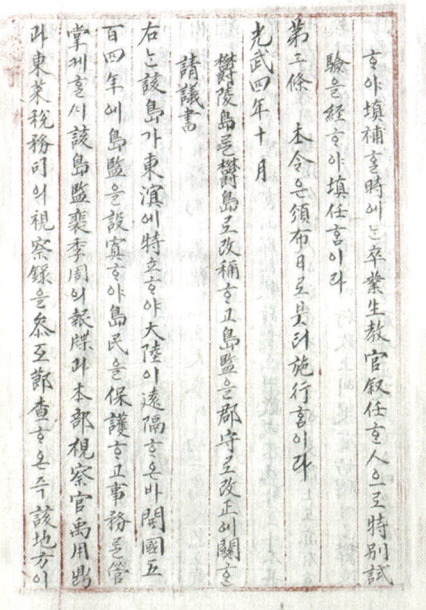

The Minister of Internal Affairs submitted a request to Kojong for a reform of Ulleungdo's administration based on Woo Yong-jeong's report.

Spring(Provided by Ulleung County Office)

24

The Russo-Japanese War and the Dokdo Watchtower

Dokdo, the First Victim to Japanese Aggression

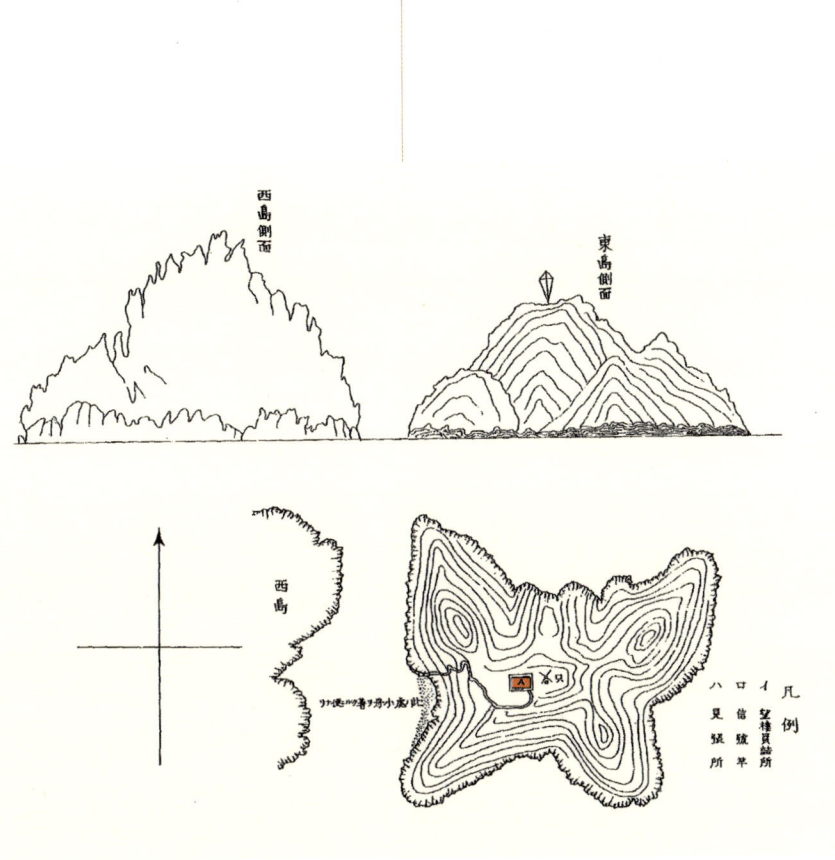

Dokdo Watchtower (Source: Japan Center for Asian Historical Recored, National Archives of Japan)

In the 20th century, as Japan planned to assume a monopoly in colonizing Korea and Russia aspired to advance to the Far East, a clash between these two powers was inevitable. Japan, realizing that she was inferior to Russia in every aspect, including military prowess, wished to resolve the problem via negotiations. However, despite Japan's few attempts to negotiate, every attempt ended in a failure.

Japan's primary demand was that Korea be declared as a protectorate, while Russia wished to protect her interests in Manchuria and prohibit Japan from using the Korean Peninsula for military purposes. The two countries could not mutually accept their demands and remained in tight opposition against each other.

In the Gozenkaigi, or "Imperial Conference," of June 3, 1903, Japan decided that even if Russia would not withdraw her forces in Manchuria and Japan had to make some concessions, Japan had to solve the "Korean Question," which was left without a solution for many years. In addition, the Diet declared in December that Korea would be "put under Japan at all costs."

With an anti-Russian Britain's support secured through the Anglo-Japanese Alliance*, Japan began the Russo-Japanese War. On February 9, 1904, the Japanese Navy ambushed and defeated 2 battleships of the Russian Navy, which were docked in waters near Incheon, Korea,

---

* A treaty signed in 1902 between Japan and Britain, which sought to prevent Russia, their common enemy, from advancing south and simultaneously protect Japanese and British interests in East Asia.

and the Japanese Army also landed in Incheon and soon entered Seoul. Furthermore, Japan and Korea signed the *Korea-Japan Protocol*, which included terms such as allowing Japan to use any region of Korea deemed necessary for Japan's military, under an austere ambience in which the Japanese exerted direct pressure through a display of threats from their military. This protocol served as the starting point for Japan's consolidation of Korea into her colony.

However, from May 15, 1904, the Japanese Navy, which had succeeded in taking the upper-hand since the beginning of the war, incurred fatal losses. The Japanese Ministry of the Navy sought to quickly occupy and transform Ulleungdo as a military base and install watchtowers aimed at discovering Russian fleets heading south and thereby supplement the numerical inferiority of her naval vessels. As tensions rose with the appearance of Russia's Pacific Fleet, the Japanese Navy wished to observe the Russian Navy's movements, and, therefore, in accordance with the *Korea-Japan Protocol*, built watchtowers in Southeastern and Northwestern Ulleungdo and began operations on September 2.

However, upon discovering that Russia's Pacific Fleet had recovered during the construction of the watchtowers and had made the Korean Strait into an enclosure, and to put up a better resistance against the Baltic Fleet, which was advancing to the East Sea, the Japanese judged that Ulleungdo's watchtower was insufficient. Hence, the Japanese decided to set up another watchtower on Dokdo, and deployed the battleship *Nitaka* to investigate the island. In short, the Japanese government was well aware of Dokdo's strategic value even before Nakai Yozaburo submitted a statement of request to annex Dokdo and in the early years of the Russo-Japanese War.

On September 29, 1904, five days after the *Nitaka* left for Dokdo to conduct an *in situ* investigation, Nakai sent a statement of request entitled *A Request for the Annexation and Lease of Dokdo* to the Home Ministry's Geography Department. Around this time, an undersea telegraph line was already established at Ulleungdo, and specific plans to extend the telegraph line to Dokdo were already prepared. However, the Home Ministry refused to accept Nakai's statement, stating that it would raise the Western Powers' suspicion that Japan had an ambition to absorb Korea.

Nakai then went to see Yamaza Enjiro, Director of Government Affairs at Japan's Ministry of Foreign Affairs. Yamaza's response was completely different from that of the Home Ministry. Yamaza ignored the Home Ministry's opinion that Dokdo 'might be Korean territory,' and led efforts to annex Dokdo in order to claim victory in the Russo-Japanese War. Yamaza hurried with annexing Dokdo because he believed that "constructing watchtowers and setting up wireless or undersea telegraph communications could be beneficial in observing the movements of enemy vessels," and "annexing Dokdo was essential under current circumstances." "Current circumstances" specifically referred to the Japanese Navy's imminent clash with the Russian Baltic Fleet, and Japan had to quickly devise measures for Dokdo as she had done for Ulleungdo. Yamaza thereby played a decisive role in annexing Dokdo.

On January 28, 1905, the Japanese Diet voted to annex Dokdo, and on

February 22, Shimane Prefecture declared the incorporation of Dokdo through Bulletin No. 40. This was a preparatory measure in response to the Baltic Fleet's invasion of the East Sea and to the transformation of the Russo-Japanese War into a protracted war. After engaging in a fierce battle with the Russian Navy on the Korean Strait on May 27, 1905, the Japanese constructed a watchtower in Dokdo on August 19.

Japan prepared an invasion of Korea in stages prior to the outbreak of the Russo-Japanese War and Dokdo became the invasion's first victim.

Summer(Provided by Ulleung County Office)

25

Acting Provincial Governor, Kangwon Province, *The Lee Myeong-rae Report*

The Korean Government's Response to Japanese Intrusions into Dokdo

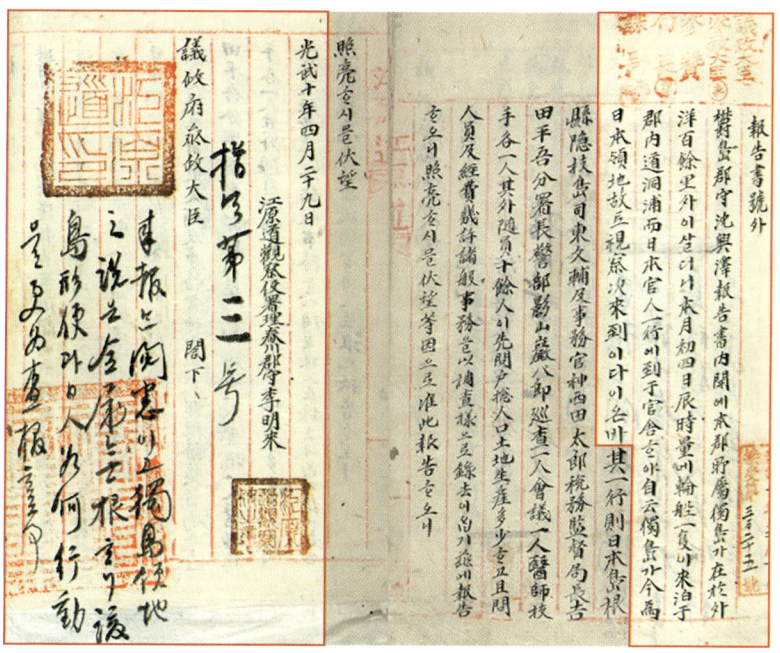

The Lee Myeong-rae Report (Source: Kyujanggak Institute for Korean Studies)

Live Update from a Report

Uldo County Head Shim Heung-taek reports that Dokdo, which is under Kangwon Province's administration, is located at about 100 li on the sea, had a steamship docked near Do-dong Port in Uldo County at about 7-9 AM, and some Japanese officials came and told him that they came to observe Dokdo, as it was declared as Japanese territory..."

April 29, Tenth Year of Kwangmu (1906)

Lee Myeong-rae, Head of Chuncheon County and Acting Provincial Governor of Kangwon Province

Directive No. 3 to His Excellency the Prime Minister

I have read the report from the acting Provincial Governor. There is no basis with which Dokdo can be declared [Japan's] territory; please observe the island's conditions and the activities of the Japanese and report back to me again.

The Lee Myeong-rae Report contains a report from Shim Heung-taek to the central government informing through Lee Myeong-rae that the Japanese had invaded Dokdo, and Prime Minister Park Je-sun's reply.

When Japanese officials stopped by Ulleungdo and told Uldo County Head Shim Heung-taek that Dokdo had been annexed by Japan, Shim informed Lee Myeong-rae, Head of Chuncheon County and Acting Provincial Governor. On April 29, 1906, Lee Myeong-rae informed Prime Minister Park Je-sun. (Right side of the photograph to the left) Consequently, on May 20, Park Je-sun, upon examining Lee's report, ordered Lee to report back to him about Dokdo's conditions and the activities of the Japanese, as there was no basis with which Dokdo could become Japanese territory. (Left side of the photograph)

Among the Korean Empire's records, the name "Dokdo" appears for the first time in this particular document. "Dokdo, which is under Kangwon Province's administration" is an important phrase which notes that Dokdo is part of Ulleungdo and is Korean territory. This report also notes with great significance that Japanese officials visited Ulleungdo after Japan illegally annexed Dokdo, and the Korean Empire clearly affirmed Korean sovereignty over Dokdo.

The government of the Korean Empire could not take any diplomatic measures because it had been stripped of diplomatic rights via the Protectorate Treaty of 1905. Yet, the *Dae-han Daily News* (May 1, 1906) and *Hwang-seong Newspaper* (May 9, 1906) reported on the *Lee Myeong-rae Report* in protest.

〈Supplemental Source 11〉 *Dae-han Maeil Shinbo (The Dae-han Daily News)* (May 1, 1906)

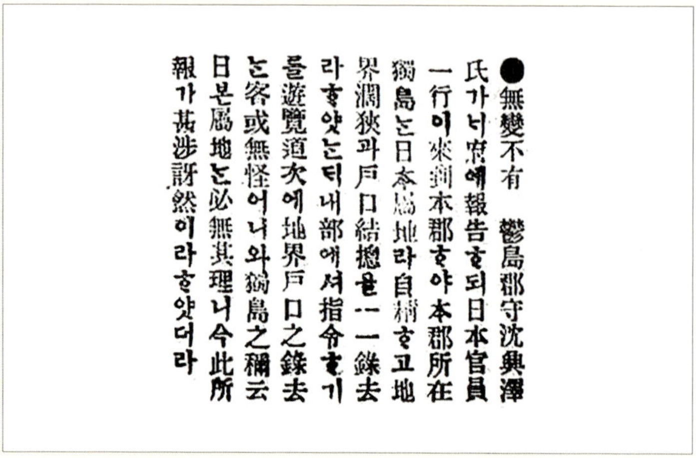

●無變不有　鬱島郡守沈興澤氏가 ᄂᆡ部에 報告ᄒᆞ되 日本官員一行이 來到本郡ᄒᆞ야 本郡所在獨島ᄂᆞᆫ 日本屬地라 自稱ᄒᆞ고 地界濶狹과 戶口結摠을 一一錄去라ᄒᆞ얏ᄂᆞᆫᄃᆡ 내部에셔 指令ᄒᆞ기를 遊覽道次에 地界戶口之錄去ᄂᆞᆫ 容或無怪어니와 獨島之稱云日본屬地ᄂᆞᆫ 必無其理니 今此所報가 甚涉訝然이라ᄒᆞ얏더라

Shim Heung-taek, Head of Uldo County, reported to the Home Ministry that Japanese officials arrived at Uldo County and declared that Dokdo was Japanese territory, and recorded the island's contours, number of households, and the sizes of every field, whereupon the Home Ministry announced that while it is possible to record the island's contours and the number of households while touring the island, and there is nothing particularly strange about such a measure, it is completely irrational to declare Dokdo as Japanese territory, and therefore, what has been recently reported is most peculiar.

Sea Chrysanthemums(Hae-Guk in Korean)
(Provided by Ulleung County Office)

The Defeat of Japan in World War II and the Return of Dokdo to Korea

26

SCAPIN No. 677 and Related Maps

The Allies Declare Dokdo To Be Outside Japan's Jurisdiction

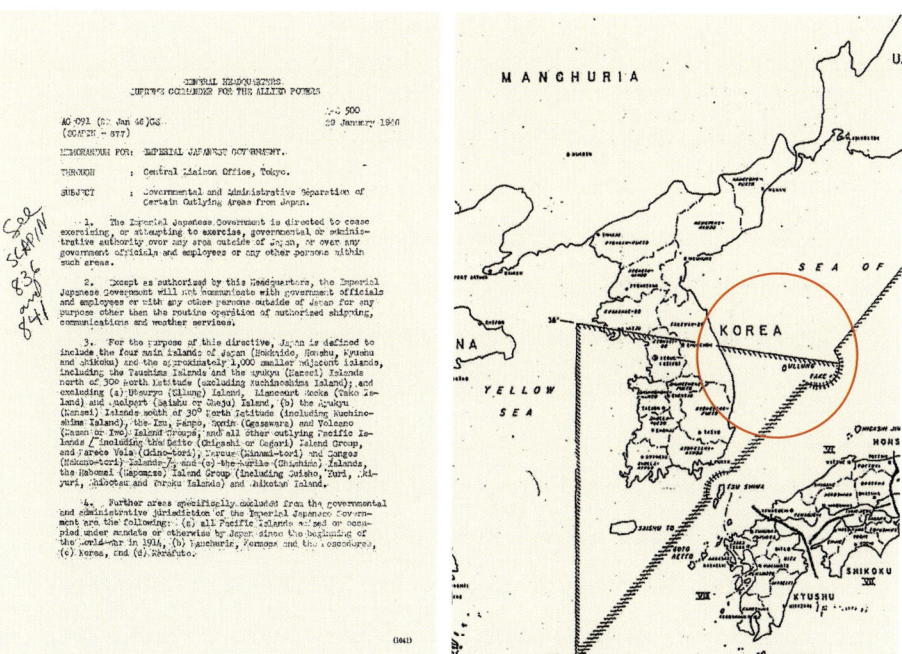

SCAPIN No. 677 and Related Maps (Source: National Archives and Records Administration, College Park, MD)

127

Memorandum for Imperial Japanese Government

Through: Central Liaison Office, Tokyo

Subject: Governmental and Administrative Separation of Certain Outlying Areas from Japan

...

3. For the purpose of this directive, Japan is defined to include the four main islands of Japan (Hokkaido, Honshu, Kyushu and Shikoku) and the approximately 1,000 smaller adjacent islands, including the Tsushima Islands and the Ryukyu (Nansei) Islands north of 30°North Latitude (excluding Kuchinoshima Island), and excluding (a) Utsryo (Ullung) Island, Liancourt Rocks (Take Island) and Quelpart (Saishu or Cheju Island), (b) the Ryukyu (Nansei) Islands south of 30°North Latitude (including Kuchinoshima Island).

This document is a memorandum which shows that the Allies considered Dokdo as a territory which Japan occupied by violence and greed and thereby politically and administratively separated Dokdo from Japan and returned it to Korea, reaffirming that Dokdo is Korean territory.

As the Second World War was entering its final stages, the leaders of the United States, Great Britain and China had a conference in Cairo, Egypt, from November 22 to November 26, 1943. The conference discussed post-war Allied policies toward Japan's territories and Korea's independence. The Cairo Declaration saw the leaders of the three nations reach an agreement on Japan's withdrawal from all territories which Japan seized by violence and greed. In addition, the Declaration stipulated that considering that Koreans are "in a condition of slavery, it is resolved that Korea must be a free and independent nation." Article 8 of the Potsdam Declaration (1945) stipulated that the terms of the Cairo Declaration shall be carried out and Japanese sovereignty shall be limited to the islands of Honshu, Hokkaido, Kyushu, Shikoku and such minor islands as we determine.

After Japan signed her declaration of surrender to the Allies on September 2, 1945, the General Headquarters was established in Tokyo and began to implement the stipulations from the Potsdam Declaration.

The Allies investigated which islands constituted "such minor islands as we determine," and on January 29, 1946, announced and implemented SCAPIN No. 677, "Governmental and Administrative Separation of Certain Outlying Areas from Japan." Article 3 defined the parameters of Japan's jurisdiction as "the four main islands of Japan—Honshu, Hokkaido, Kyushu, and Shikoku and approximately 1,000 adjacent islands." Article 3 then mentioned Ulleungdo, Jejudo and Dokdo as islands excluded from Japan's territory.

The Allies announced this after reaching a decision following investigations which took several months, and as the Supreme Commander of the Allied Powers was, in accordance with contemporary international law, a legal institution, the SCAP's decision to declare Dokdo as Korean territory by returning it to its genuine owner has validity under international law. This decision is evidently marked on the SCAP's map for administrative jurisdiction.

The Allies' decision to separate Dokdo from Japan's territory is in accordance with the Allies' post-war policy affirmed by the Cairo Declaration and the Potsdam Declaration's demand that Japan must forfeit territories "seized by violence and greed." Thus, Dokdo, which was seized by violence and greed in the process of colonizing Korea and during the Russo-Japanese War, was a territory from which Japan must withdraw insofar as she accepted the terms of the Potsdam Declaration upon her unconditional surrender.

Japan claims that Dokdo was not originally Korean territory, and that the territories which Japan returned to Korea after World War II were not seized by violence and greed, but were Korean territories when Japan annexed Korea in August 1910. Japan argues that therefore, Dokdo cannot be classified as a "region which Japan seized by violence and greed."

Japan also takes issue with Article 6 of SCAPIN No. 677, which states that "nothing in this directive shall be construed as an indication of Allied policy relating to the ultimate determination of the minor islands referred to in Article 8 of the Potsdam Declaration." In other words, the Article mentions that there is no "ultimate decision." After initiating a debate on the issue of territorial sovereignty over Dokdo in 1952, the Japanese government sent a statement on April 25, which claimed that per Article 6, Japan's territories were never ultimately determined.

However, what is emphasized in SCAPIN No. 677 is that considering the possibility of an Ally raising an objection amidst the complex web of interests among Allied nations, was not that there would be any "ultimate decision", but rather that there would be no Allied policy on an ultimate decision. This only implies the possibility of revising the ultimate decision, if there are any objections from any Allied nation in the future,.

So what happens when necessary revisions have to be made? Article 5 of SCAPIN No. 677 states that, "the definition of Japan contained in this directive shall also apply to all future directives, memorandum and orders from this Headquarters unless otherwise specified therein." Therefore, should there be any need to revise the territorial definition of "Japan," the revision can occur through directives, memorandum, and orders. The Article clearly notes that unless revisions are required, the territorial definition of "Japan" applies to the future.

In other words, if Article 5 of SCAPIN No. 677 is applied to Dokdo, when Ulleungdo and Jejudo are separated from Japan per Article 3 but Japan's return of Dokdo to Korea as Korean territory needs revision per Article 5, the revision can only come into effect if SCAP issues a different directive, memorandum, or order stating that "Dokdo, which had been returned to Korea, is to be incorporated as part of Japan on this occasion."

From its initial announcement of SCAPIN No. 677 until its dissolution in 1952, the General Headquarters of the Allied Powers never issued a different directive, memorandum, or order stating that Dokdo will be incorporated as a part of Japan. Hence, SCAPIN No. 677 reaffirms Dokdo as Korean territory under international law, and continues to legally be occupied by Korea in accordance with international law.

The Japanese government also raises an objection to this issue, claiming that the San Francisco Peace Treaty is the only legally effective treaty regarding Dokdo. Article 2 (a) does not clearly specify to whom Dokdo belongs. Since the Treaty does not mention Dokdo, the Japanese claim that it is Japanese territory. The Treaty lists only three islands (Jejudo, Ulleungdo, and Geomundo), so according to Japan's logic, all islands associated with the Korean Peninsula can become Japanese territory. Since it is impossible to list all islands associated with the Korean Peninsula, which number in the thousands, how can the omission of all their names in the Treaty become evidence that they are all Japan's territories? To the contrary, Article 8 of the San Francisco Peace Treaty states that Japan will recognize the "full force of all treaties now or hereafter concluded by the Allied Powers for terminating the state of war initiated on September 1, 1939, as well as any other arrangements by the Allied Powers for or in connection with the restoration of peace," thereby affirming the sustained legal validity of SCAPIN No. 677.

In short, since SCAPIN No. 677 excluded Dokdo from Japan's administrative regions and there are no measures described in the San Francisco Peace Treaty which stipulate that Dokdo is Japanese territory, Japan's claim that it was "the Allies' or the United States' intention to leave Dokdo as Japan's territory" is unsound.

27
Investigation Commission for Scholarly Research on Ulleungdo and Dokdo
Korea's Defense of Dokdo after Liberation

1947. 8.

1953

Investigation Commission for Scholarly Research on Ulleungdo and Dokdo
(Source: Ministry of Foreign Affairs of the Republic of Korea)

In early August 1947, the Provisional Government of the Republic of Korea, under the auspices of the American Military Government organized an "Investigative Committee on Dokdo," with An Jae-hong, Korea's Minister of Civil Affairs as the Committee's chairman. On August 4, after opening a joint conference with officials of Central Administrative Office and experts, the Committee decided to deploy an Investigation Commission for Dokdo with the aim of discovering historical documents and surveying Dokdo. Accordingly, a two-week plan which would begin on August 16 and end on August 28 was set up, with the Korean Alpine Association* Serving as the host and the Ministry of Education as a sponsor. The first Investigation Commission for Scholarly Research on Ulleungdo and Dokdo was formed through a joint civilian-official cooperation.

The team's members included Shin Seok-ho, Head Curator of the National History Museum, Chu In-bong, Chief of the Japan Desk at the Ministry of Foreign Affairs, Lee Bong-soo Chief Editor of the Ministry of Education, and Han Ki-jun, an engineer at the Fisheries Department as representatives from the Provisional Government, 63 experts from various fields represented by Song Seok-ha, chairman of the Korean Alpine Association, 2 officials from the Central Administrative Office of North Kyeong-sang Province, and police officers from the 5th Precinct, for a total of 80 people. The investigation teams were massive. After conducting *in situ* research at Dokdo, the teams established two

---

* It was established on September 15, 1945. The Team would change its name to the Korean Alpine Association on the occasion of the founding of the Republic of Korea's government.

wooden signposts which first informed that Dokdo is Korean territory. The signpost on the right read "朝鮮 鬱陵島 南面 獨島" (Dokdo, South County, Ulleungdo, Korea) and the signpost on the left read "鬱陵島, 學術調査隊 紀念" (Ulleungdo, in honor of the Investigation Commission for Scholarly Research). The signposts were established on August 20, 1947.

Judging from the fact that the Committee's chairman was a Minister of Civil Affairs in the Provisional Government, that the Committee, under the agreement of both relevant institutions and experts, decided to deploy investigation teams and organized and arranged them, and finally, that the investigation teams used a 300-ton small naval vessel called the Dae-jeon, which was affiliated with the Coast Guard of Korea, tells us that the first investigative teams were part of an official investigative activity which was authorized by the government and had won the government's support.

In September, 1952, the Second Investigation Commission for Scholarly Research on Ulleungdo and Dokdo was deployed with the Korean Alpine Association once again serving as the host. There were 38 people in this team and Hong Jong-in was its captain.

This investigation was also supported by the Ministry of Education, the Ministry of Foreign Affairs, the Ministry of Defense, the Ministry of Commerce and Industry, and the Ministry of Public Security as a pan-governmental effort. The investigation team boarded the Jin-nam, a lighthouse cruise ship operated by the Busan Ministry of Transportation's Maritime Affairs Division, on September 17, five days later than the original schedule due to typhoons, and arrived at Ulleungdo's Do-dong Port the next day. When the investigation team was about 2 KM in front Dokdo at 11 AM on September 22 after departing from Do-dong Port at 5:30 AM on September 18, 4 planes from the U.S. Air Force were conducting mock-bombing raids, and the ship had to return to Ulleungdo. The investigation team resumed its investigation and advance to about 1 KM in front of Dokdo, but 2-4 twin bomber planes once again dropped about a dozen bombs on Dokdo, hindering the team from landing on Dokdo, whereupon the team had to return to Busan.

The investigative report of Park Byeong-ju of the Surveying and Geodesic Department reported the following:

> The investigation team arrived at waters nearby Dokdo, braving through strong winds and waves. 3 military planes suddenly appeared and, ignoring the *Jin-nam's* presence, continued to bomb Dokdo, and we failed to land on the island, whereupon we barely managed to measure the height of Dokdo's West Island, which was about 130 meters, by using sight, sound, and angular measurements. The team approached Dokdo for the second time on September 24, but the waves were strong and the bombings started again, so we navigated around the island's parameters and just took photographs and retreated.

One year later, on July 8, 1953, the National Assembly passed a resolution during the 19th Central Meeting to "support the deployment of a qualitatively improved investigation team which includes the South Korean Alpine Association. In accordance with this resolution, the government encouraged the South Korean Alpine Association (formerly the Korean Alpine Association) to organize and deploy the Third Investigation Commission. The South Korean Alpine Association devised a plan on two occasions from July to September 1953 to deploy the Third Investigation Commission for Scholarly Research on Ulleungdo and Dokdo. Under the auspices of various government agencies, 38 members, 3 officials from the Central Administrative Office of North Kyeong-sang Province, and 20 officials related to Ulleungdo formed an investigative team. Hong Jong-in once again served as the team's leader.

Among the objectives for the deployment of the Investigation Commission were activities which were not part of previous investigation plans, or 1) scientific investigation of waters near Dokdo (geology, climate, conditions of the sea, organisms, marine life, history, geography) and 2) a survey of Dokdo and making cartographic records. The Investigation Commission boarded Naval Ship No. 905 and investigated and surveyed Dokdo on October 14 and 15, or for two days.

On October 15, 1953, the Investigation Commission placed a granite plaque made on August 15, 1952. (Confer Supplemental Source 12) The Second Investigation Commission had previously failed to erect a granite plaque due to the bombing sessions on Dokdo. The front of the plaque reads "Dokdo, 獨島, LIANCOURT ROCKS," and at the back of the plaque are the words, "Korean Alpine Association (in Korean) Ulleungdo Dokdo Investigation Commission for Scholarly Research Korea Alpine Association 15th Aug. 1952." The current granite plaque was reconstructed and established by the South Korean Alpine Association on July 6, 2015 after receiving the government's permission to change the form of a national cultural landmark.

⟨**Supplemental Source 12**⟩ The Front and Back of the Plaque Erected in 1953

Front Captain Heung Jong-in of the Investigation Commission for Scholarly Research on Ulleungdo and Dokdo, removes pickets installed by the Japanese and places the plaque. LIANCOURT is engraved on the plaque.

Back October 15, 1953, the actual day in which the plaque was erected was additionally inscribed later.

⟨**Supplemental Source 13**⟩ The plaque, reinstalled on July 6, 2015

A Panoramic view of West Island as seen from East Island's Mongdol Coast, site of the Dokdo granite plaque.

Black-tailed Gulls(Provided by Ulleung County Office)

28
The Dokdo Bombing Incident (1948, 1952)
Dokdo Remains as a Home for the Korean People Despite Adversities

Memorial Monument for Victimized Fishermen

A memorial monument for victimized fishermen is established on Dokdo's East Island coast. It was established in 1950 for fishermen who were victims to mock-bombing sessions carried out by the U.S. Air Force on June 8, 1948. The current monument is a replica of the original version, and was established in 2005. It is uncertain why the original monument disappeared, whether it was due to a typhoon or by a particular individual. The original monument was discovered in the sea near Dokdo, and is now relocated and preserved at Ulleungdo's An Yong-bok Memorial Museum.

The Supreme Commander of the Allied Powers in Japan excluded Dokdo from Japan's territorial jurisdiction and prohibited the Japanese from fishing near Dokdo's waters by announcing SCAP No. 677 and No. 1033 (Confer Source 26). By contrast, since Dokdo was under Korean control, Korean fishermen were allowed to fish near the island. However, on September 16, 1947, the Supreme Commander of the Allied Powers designated Dokdo as a training ground for aerial bombing practice drills.

On June 8, 1948, Korean fishermen had gathered at Dokdo from Ulleungdo and Kangwon Province and were peacefully collecting seaweed. At around 12 PM, American military planes based in Okinawa dropped dozens of bombs as part of their practice drill. 14 fishermen were killed, many were wounded, and many fishing vessels were destroyed as a result of the bombing.

The Korean media thoroughly reported on this incident, and American newspapers also paid attention and covered it. *The New York Times* particularly noted that "Dokdo is a region from which Korean fishermen had made their livelihoods for generations," and emphasized the need for a quick solution to the problem. (June 17, 1948) In addition, William Dean, Minister of Military Affairs in the American military government in Korea, requested to the Commander of the U.S. Far

Dong-a Il-bo(East Asia Daily) article of September 21, 1952

Eastern Forces in Japan that because "the waters near Dokdo are most ideal for Korean fishermen to engage in fishing," bombing drills had to be ceased. (Telegram from June 24, 1948)

The U.S. Forces in Korea sent an investigation team to Dokdo to inspect the incident, and compensated for the loss of human life and damages done to the fishing vessels. It was a tragic incident featuring the sacrifice of many fishermen, but this incident demonstrates that Dokdo was under Korean jurisdiction after Korea's liberation, and that the island was essential for Korean fishermen to lead their lives.

On June 8, 1950, the opening ceremony and a performance of rituals for the victimized fishermen took place, with the Governor of North Kyeongsang Province (Kim Jae-cheon) in attendance. However, after a little more than 2 years later, on September 15, 1952, the American Air Force carried out another aerial bombing practice drill on Dokdo. Several Korean fishermen had to be forced to escape to Dokdo, but fortunately did so unscathed. Yet, on September 22 and September 24, 1952, American planes dropped bombs whenever the Investigation Commission for Scholarly Research on Dokdo approached Dokdo's coast.

The Korean government inquired the American military for facts regarding the bombing practice drills and learned that Dokdo was once again designated as a training ground for aerial bombing sessions in July 1952. What proved even more surprising was Japan's involvement in these practice drills. When the Korean government declared a line of peace including Dokdo in Korea's maritime space, the Japanese government argued for Japan's sovereignty over Dokdo and tried to nullify the line of peace. Japan sought to induce the Americans to use Dokdo as a training ground for aerial bombing drills as one method of nullification. However, the Japanese government's plan came to nothing. After the second aerial bombing practice drill, the United States received protests from the Korean government and excluded Dokdo from being a training ground for aerial bombing practice drills.

After Dokdo's exclusion from the aerial bombing practice drills, the Korean Alpine Association organized another scholarly investigation of Dokdo and established the territorial plaque on Dokdo's East Coast, as it had been originally planned in 1952. (Confer Source 27)

29

The Declaration of a Line of Peace, Gazette, and Map(1952)

A Reconfirmation of Dokdo's Identity as a Territory of the Republic of Korea to the World

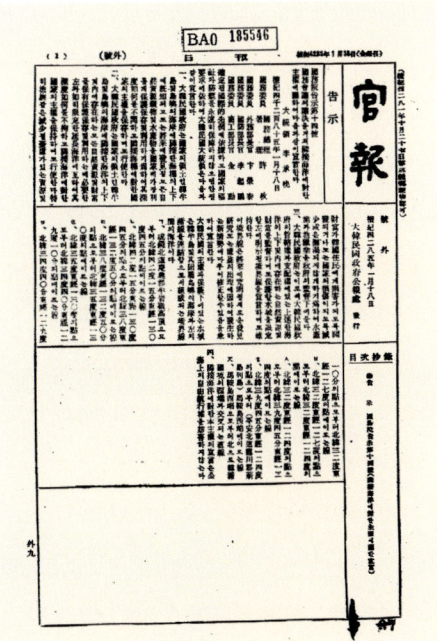

A Public Notice Gazette Containing
the Declaration of a Line of Peace

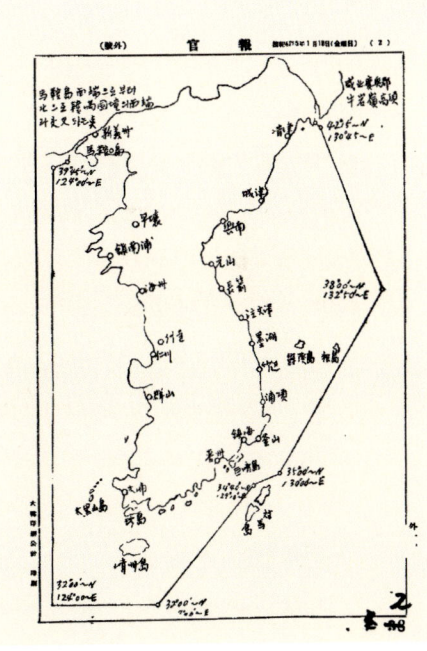

Map Accompanying the Gazette

The gazette and accompanying map, as shown above, places Dokdo within the parameters of the Line of Peace, illustrating that Dokdo is an island within Korea's sovereign jurisdiction and publicizes the Declaration of a Line of Peace. On January 18, 1952, while the Korean War was still ongoing, the South Korean government issued a Presidential Statement on Korea's Sovereignty over Adjacent Waters. (Public Notice No. 14 of the Department of National Affairs, also known as the Declaration of a Line of Peace) In 1952, the MacArthur Line was in effect in the waters between Korea and Japan. The MacArthur Line was established in September 1945 by the Allies to limit Japanese fishing activities. However, in April 1952, it became clear that once the San Francisco Peace Treaty would come into effect, the MacArthur Line would be nullified. Once the MacArthur Line was nullified, it was evident that Japanese fishing vessels would swarm over to Korea's coasts. The South Korean government declared a Line of Peace in preparation for such a scenario.

The Declaration was publicly announced by the President after passing the National Affairs Meeting's resolution and consisted of four articles. The Declaration announced Korean sovereignty over the Korean Peninsula and fishing rights and underwater mineral resources in seas near islands surrounding the Peninsula, and marked the borders of these seas. In addition, by placing Dokdo inside the Line of Peace, the government clearly noted that Dokdo is an island within South Korean sovereignty.

Japan showed vehement opposition once the Declaration of the Line of Peace was announced. Japan argued that the Line of Peace infringed the freedom of the seas under international law and curtailed her sovereignty over Dokdo. Japan sought to nullify the Line of Peace using all necessary means.

Since Korea continuously occupied Dokdo, Japan's argument for her territorial sovereignty over Dokdo was outrageous. Moreover, should the MacArthur Line be abolished amidst the chaos caused by the Korean War, Korea expected Japan's indiscriminate overfishing and sought to protect Korea's marine resources, for which the Declaration was an urgent yet fundamental measure. The Declaration inherited the MacArthur Line and guaranteed the freedom of navigation and reflected new trends in international law, as shown in President Truman's "Policy of the United States with Respect to the Natural Resources of the Subsoil and Sea Bed of the Continental Shelf."

「Timeless Moonlit Nights」(KimJungman, 2014)

30

The Dokdo Volunteer Garrison and the Dokdo Security Police

Defending Dokdo for the Korean People

The Dokdo Volunteer Garrison inspecting with a large telescope.

The Dokdo Security Police Raising the Republic of Korea's National Flag

In March 1953, after Dokdo became excluded from being a training ground for the American Air Force's bombing practice drills, there were instances of Japanese illegally landing on Dokdo or putting up signposts reading "Takeshima is Japanese Territory." During these incidents, several Ulleungdo fishermen who had visited Dokdo for fishing activities were forced to turn away by the Japanese.

As such incidents became frequent, Ulleungdo residents established the Dokdo Volunteer Garrison to prevent illegal Japanese intrusions into Dokdo and protect the livelihoods of Ulleungdo residents. The Garrison had many achievements such as repelling Japanese patrol ships which had advanced towards Dokdo in August and November 1954, playing a decisive role in thwarting Japanese ambitions to invade the island.

Since the announcement of the Line of Peace in 1952, the South Korean government also devised methods to defend Dokdo from Japan's invasive activities. The Ulleung Police Station formed and operated a team of Dokdo patrol guards, and the government of the Republic of Korea decided to establish a permanent guard post at Dokdo. The Korean government deployed police forces to Dokdo and specially employed the Dokdo Volunteer Garrison as police officers to strengthen the Dokdo Security Police.

The Dokdo Security Police currently maintains one platoon to prepare for possible intrusions from Japanese patrol ships or other foreign forces and is on guard 24 hours a day along the coast. The Dokdo Security Police not only possesses patrolling vessels from the National Maritime Police Force but also telecommunication facilities to communicate with the Korean Navy and Air Force, thereby maintaining a trustworthy defense of Dokdo.

INDEX

• A •

A Complete Map of Ezo 56
A Complete Map of Korea 65, 69, 77, 78, 66, 70
A Complete Map of the Korean Kingdom 45~48
A Complete Map of the Korean Nation 56, 57
A Complete Map of Ryukyu 56
A Comprehensive Map of the Eight Provinces 21, 22, 50
A Comprehensive Map of Our Country 59, 60
Haedongjeondo (A Complete Map of Korea) 65, 66
A Map of Isotakeshima 91, 92, 94, 95
Admiral Putschachin 72, 73
A Gazetteer on Korea's Waterways 89, 90
A General Cartographic Record of Japan in An Illustrated Description of Three Countries 55~58
A General Introduction to Korea's Financial and Military Affairs 54
A Grand Map of Japan's Coasts and main Territories 99
A Grand Map of Three Countries—Hokkaido, Korea, and Japan 58
A History of Goryeo 16, 19
A History of the Three Kingdoms 15, 16
A List of Korea's Eight Provinces 31, 32
A Map of Korea's East Coast 87~89
A Map of Korea's Main Territories 60
A Map of Korea 61, 66, 70~75, 87~89
A Map of Three Countries Adjacent to Japan 56
A New and an Expanded Encyclopedia of Korean Geography 16, 21, 22
A Record of General Observations on Oki Island 23~25
A Request from the Minister of Internal Affairs 111
A Summary of Reasons and Causes 92, 94
An Encyclopedia of Korean Customs and Culture 31, 53, 54

An Illustration of an Uninhabited Island
   56
Anglo-Japanese Alliance   114
An Investigative Diary on Ulleungdo
   103, 104
An Yong-bok   30~34, 51, 54, 60, 61,
   142
An Yong-bok Memorial Museum   142
An Encyclopedia of Korean
   Geography   53
A Geographical Survey of Korea in The
   Royal Annals of King Sejong   10, 17,
   19
An Extended Encyclopedia of Korean
   Customs and Culture   54
An Illustrated Description of Three
   Countries   55~58
An Intelligence Report on the Internal
   Affairs of Korea   81, 82
An Order Prohibiting the Crossing of
   the Seas   36, 94
Argonaut Island   74

• B •

Bamboo Island   40, 108, 110

Baekdu Border Monument   79
Bang Ji-yong   20
border of an area which the Qing
   sought to restrict the Han's access
   47

• C •

Cairo Declaration   128~130

• D •

Dagelet   74
J. B. B. D'Anville   45~47
Dokdo Volunteer Garrison   149, 151
Dae-han Daily News   121
Dajokan   81, 83, 86, 91, 94~96
Dajokan Directive   91
Dokdo Security Police   149~151
Dokdo watchtower   113, 115, 117
Jean Du Halde   46

• E •

East Sea   22, 66, 71~75, 88, 89, 117
encouraged to relocate to mainland

Korea 51

• G •

General Headquarters of the Allied
    Powers 131
Governor of Ulleungdo 109
granite plaque 137, 139

• H •

Hasla Province 16, 18
Hayashi Shihei 55, 56, 58
Hong Jong-in 135, 137
Hwang-seong Newspaper 121

• I •

*Imperial Decree No. 41 of the Korean
    Empire* 105, 107, 108, 110
Incorporation of Dokdo 117
Ino Tadataka 99
inscribed on a rock 104
Inspector 51, 66, 67, 104
Investigation Commission for Scholarly
    Research on Dokdo 133

Investigation Commission for Scholarly
    Research on Ulleungdo and Dokdo
    134, 135
Investigator for Ulleungdo 104
Isotakeshima 24, 57, 91~95

• J •

Jang Han-sang 10, 37, 38, 40, 51
Japan-Korea Protectorqte Treaty 121
Jasando 31
Jeong Sang-ki 50, 60, 61, 66, 70, 79

• K •

Kangwon Province 20, 31, 32, 60,
    65~67, 79, 108, 119~121, 142
Kim Bu-sik 15, 16
Kim Dae-geon 69, 70
Kim In-woo 18, 20
Kim Myeon-soo 108, 109
Kimura Nobuaki 98
Kim Yoo-rip 18
King Jijeung 15, 16, 18
King Taejong 20
Korean Alpine Association 134~137

Korea-Japan Protocol 115

• L •

Laporte 109
Lee Kyu-won 103~106
Lee Man 20
Lee Myeong-rae 119~121
Lee Myeong-rae Report 119, 120
Lee Sa-bu 16, 18, 80
Line of Peace 114, 145~147
logging 104, 105, 108
Lord of Tsushima 30

• M •

MacArthur Line 146, 147
Matsudaira Shintaro 34, 36
Matsushima 11, 24, 30, 31, 34, 35, 53, 54, 58, 82, 84, 85, 93~95
Memorial Monument for Victimized Fishermen 141
Menelai 72~74, 88, 90
Middle Peak 38, 40, 51, 80
Moriyama Shigeru 83
Mureung Island 20

• N •

Nakai Yozaburo 115
Nam Ku-man 38
New Collection of Maps of China 46, 49
Nitaka 115, 116

• O •

Oki Island 10, 11, 23~25, 31, 57, 58, 92
Olibutscha 72~75, 88, 90
Overseer 105, 108
order to develop farmlands in Ulleungdo 104

• P •

the *Palada* 72, 73, 75
Park Byeong-ju 136
Park Eo-dun 30, 34, 51
Park Je-sun 120
Park Seup 20
Potsdam Declaration 129, 130
Provincial Governor 119, 120
Public Security Manager 18, 20

• R •

Record of China's Geography and History 46
"Reflections on Korean's Territory" 54
Régis Line 47, 48
Royal Annals of King Taejong 20
Royal Examiner 18
Russo-Japanese War 113, 114, 116, 117, 130

• S •

Sada Hakubo 83
Saito Sakae 83
Saito Toyonobu 23, 24
Samcheok 10, 18, 20, 39, 40, 51, 67
sea lion 24, 35, 40
Seaside bamboos 38, 39
Seikanron 83
Shim Heung-taek Report 120
Shin Kyeong-jun 54
San Francisco Peace Treaty 131, 132, 146
Supreme Commander of the Allied Powers 129, 131

SCAPIN No. 677 131, 132
Seokdo 108, 110
Sergeiev 74, 75
Shimane Prefecture 24, 92~95, 117
Shimane Prefecture Bulletin Notice No. 40 117

• T •

Takeshima 150
Takeshima Incident 34, 85
Takeshima and one other island 93~95
Territorial sovereignty over Dokdo 108, 130, 147
The Baltic Fleet 115~117
The Grand Map of Korea 49, 50, 51
The Main Historical Record of Silla 16
The Meiji Restoration 83, 88
Three Peaks 38
Tottori 31~35, 95

• U •

"Ulleungdo Controversy" 30, 34, 38, 51, 54, 85

Usankuk   15, 16, 18, 57, 80

・ V ・

Views on Isotakeshima   94, 95

・ W ・

"Willow-Fence Border"   48
Woo Yong-jeong   108~110

・ Y ・

Yamaza Enjiro   116
Yoshida Shoin   83

Notation of Dokdo as a Territory of the Republic of Korea

The Dokdo Security Police

ⓒ KimJungman

Seeing Dokdo Through 30 Images and Historical Documents

Published on	November 5, 2020
Edited by	Dokdo Research Institute, Northeast Asian History Foundation
Published by	Northeast Asian History Foundation
Publisher	Kim Dohyung
Registration	312-2004-050(2004. 10. 18.)
Address	81, Tongil-ro, Seodaemun-gu, Seoul, Republic of Korea
Tel	02-2012-6065
Fax	02-2012-6189
Website	www.nahf.or.kr
Printed in	Yeoksagonggan

© Northeast Asian History Foundation, 2020

ISBN 978-89-6187-563-9 04910
978-89-6187-482-3 (set)